T0366024

Instituting Nature

Politics, Science, and the Environment

Peter M. Haas and Sheila Jasanoff, editors

Peter Dauvergne, *Shadows in the Forest: Japan and the Politics of Timber in Southeast Asia*

Peter Cebon, Urs Dahinden, Huw Davies, Dieter M. Imboden, and Carlo C. Jaeger, eds., *Views from the Alps: Regional Perspectives on Climate Change*

Clark C. Gibson, Margaret A. McKean, and Elinor Ostrom, eds., *People and Forests: Communities, Institutions, and Governance*

The Social Learning Group, *Learning to Manage Global Environmental Risks. Volume 1: A Comparative History of Social Responses to Climate Change, Ozone Depletion, and Acid Rain. Volume 2: A Functional Analysis of Social Responses to Climate Change, Ozone Depletion, and Acid Rain*

Clark Miller and Paul N. Edwards, eds., *Changing the Atmosphere: Expert Knowledge and Environmental Governance*

Craig W. Thomas, *Bureaucratic Landscapes: Interagency Cooperation and the Preservation of Biodiversity*

Nives Dolsak and Elinor Ostrom, eds., *The Commons in the New Millennium: Challenges and Adaptation*

Kenneth E. Wilkening, *Acid Rain Science and Politics in Japan: A History of Knowledge and Action toward Sustainability*

Virginia M. Walsh, *Global Institutions and Social Knowledge: Generating Research at the Scripps Institution and the Inter-American Tropical Tuna Commission, 1900s–1990s*

Sheila Jasanoff and Marybeth Long Martello, eds., *Earthly Politics: Local and Global in Environmental Governance*

Christopher Ansell and David Vogel, eds., *What's the Beef? The Contested Governance of European Food Safety*

Charlotte Epstein, *The Power of Words in International Relations: Birth of an Anti-Whaling Discourse*

Ann Campbell Keller, *Science in Environmental Politics: The Politics of Objective Advice*

Henrik Selin, *Global Governance of Hazardous Chemicals: Challenges of Multilevel Management*

Rolf Lidskog and Göran Sundqvist, eds., *Governing the Air: The Dynamics of Science, Policy, and-Citizen Interaction*

Andrew S. Mathews, *Instituting Nature: Authority, Expertise, and Power in Mexican Forests*

Instituting Nature

Authority, Expertise, and Power in Mexican Forests

Andrew S. Mathews

The MIT Press
Cambridge, Massachusetts
London, England

This book was set in Sabon by Toppan Best-set Premedia Limited.

Library of Congress Cataloging-in-Publication Data

Mathews, Andrew S.
Instituting nature : authority, expertise, and power in Mexican forests / Andrew S. Mathews.
p. cm. — (Politics, science, and the environment)
Includes bibliographical references and index.
ISBN 978-0-262-01652-0 (hardcover : alk. paper) -- ISBN 978-0-262-51644-0 (pbk. : alk. paper)
1. Zapotec Indians—Mexico—Ixtlán de Juárez--Social conditions. 2. Zapotec Indians—Mexico—Ixtlán de Juárez—Industries. 3. Zapotec Indians—Mexico—Ixtlán de Juárez—Government relations. 4. Indigenous peoples—Ecology—Mexico—Ixtlán de Juárez. 5. Forests and forestry—Mexico—Ixtlán de Juárez. 6. Forest management—Mexico—Ixtlán de Juárez. 7. Forest conservation—Mexico—Ixtlán de Juárez. 8. Ixtlán de Juárez (Mexico)—Politics and government. 9. Ixtlán de Juárez (Mexico)—Social conditions. 10. Ixtlán de Juárez (Mexico)—Environmental conditions. I. Title.
F1221.Z3M373 2011
333.75'16097274—dc22

2011009095

Contents

Series Foreword vii
Acknowledgments ix
Glossary of Institutions xi

1 Introduction 1

2 Building Forestry in Mexico: Ambitious Regulations and Popular Evasions 31

3 The Sierra Juárez of Oaxaca: Mobile Landscapes, Political Economy, and the Fires of War 61

4 Forestry Comes to Oaxaca: Bureaucrats, Gangsters, and Indigenous Communities, 1926–1956 93

5 Industrial Forestry, Watershed Control, and the Rise of Community Forestry, 1956–2001 117

6 The Mexican Forest Service: Knowledge, Ignorance, and Power 147

7 The Acrobatics of Transparency and Obscurity: Forestry Regulations Travel to Oaxaca 179

8 Working the Indigenous Industrial 203

9 Conclusion 235

Appendix 243
Notes 251
References 267
Index 291

Series Foreword

As our understanding of environmental threats deepens and broadens, it is increasingly clear that many environmental issues cannot be simply understood, analyzed, or acted on. The multifaceted relationships among human beings, social and political institutions, and the physical environment in which they are situated extend across disciplinary as well as geopolitical confines and cannot be analyzed or resolved in isolation.

The purpose of this series is to address the increasingly complex questions of how societies come to understand, confront, and cope with both the sources and manifestations of present and potential environmental threats. Works in the series may focus on matters political, scientific, technical, social, or economic. What they share is attention to the intertwined roles of politics, science, and technology in the recognition, framing, analysis, and management of environmentally related contemporary issues and a manifest relevance to the increasingly difficult problems of identifying and forging environmentally sound public policy.

Peter M. Haas
Sheila Jasanoff

Acknowledgments

Anthropologists know that debts can be powerful and that they can establish enduring social relationships. If this is so, I am rich in the many debts that I have incurred in writing this book. Although I alone am responsible for errors and mistakes, I have been influenced by many people, and this book is part of a long conversation with them.

This book project began during my doctoral study, and I must acknowledge the intellectual influence of Michael R. Dove, who encouraged my initial ideas and helped me think that ignorance could be a form of power, without ever imposing on me his own view of the world. The other great influence has been Sheila Jasanoff, who introduced me to the worlds of Science and Technology Studies, helping me to step sideways and see my project from a different direction entirely. When I first went to Yale, K. Shivaramakrishnan encouraged me to think with anthropology, and Enrique Mayer and David Graeber helped me to become a practicing anthropologist; Mark Ashton taught me forest ecology and silviculture, and Ann Camp taught me how to study tree rings. Richard Tardanico at Florida International University (FIU) helped organize a research leave to work on the book manuscript: I was fortunate indeed in my colleagues at FIU, including Gail Hollander, Laura Ogden, and Rebecca Zarger. David Bray, who read drafts of chapter 7 and what are now parts of chapters 3 and 4, was the first of these scholars to welcome me. At FIU, Rod Neumann read a draft of chapter 7, and Laura Ogden read versions of chapters 6 and 7. At the University of California (UC) Santa Cruz, I have been blessed with tough-minded colleagues who are passionate about writing: Melissa Caldwell read drafts of chapters 6 and 7, and Anna Tsing read a draft of chapter 1. Their advice and insights have helped me greatly. It is customary to thank manuscript reviewers, but in this case my thanks are more than a perfunctory acknowledgment: The reviewers convened by the MIT Press helped me tighten and enliven the argument greatly, as did advice from Sheila Jasanoff and Clay Morgan.

Many other people have helped me through their responses to chapters or conference papers that touch on parts of this work. My thanks to Jonathan Padwe, Laura Meitzner Yoder, Steve Rhee, Anne Rademacher, Julie Chu, Myanna Lahsen, Antonio Azuela, and Guadalupe Rodriguez Gomez. In Mexico I have many debts indeed: First and foremost, I must thank the *comuneros* and authorities of Ixtlán de Juárez (most of whom remain un-named), but especially Pedro Vidal Garcia Pérez, Leopoldo Santiago Pérez, and Gustavo Ramirez Santiago. My thanks also to Martin Gomez Cardenas and Juan Francisco Castellanos at INIFAP in Oaxaca, Antonio Plancarte at SEMARNAP Oaxaca, and Jose Luis Romo at UNAM Chapingo, as well as Aurelio Martin Fierros. I must also thank many unnamed forestry officials and foresters, who took time out of busy lives to attend to my questions.

This book has been supported by many funders over a number of years. The research on which this book is based was supported by the Tropical Resource Institute of the Yale School of Forestry and Environmental Studies, by an Enders Grant from the Yale Graduate School of Arts and Sciences, by the Yale Center for International and Area Studies, by a Switzer Fellowship, by a Fulbright/García-Robles fellowship, and by a grant from the National Science Foundation Program in Science and Technology Studies (no. BR-0002132). A teaching leave from FIU helped greatly during the writing, while additional grant support from the UC Santa Cruz Committee on Research and from UC MEXUS-CONACYT supported fieldwork in the summers of 2008 and 2009. I thank all of these supporters.

Finally, I must thank my friends and family: In Mexico, the Prieto family supported my research from its earliest days, putting me up on my travels and welcoming me into their homes. My family has lived for a long time with this project: This book is especially dedicated to my grandmother, Nieves de Madariaga Mathews, who died before this project was complete, and also to her sister, Isabel de Madariaga. I have been fortunate to grow up with people who think that arguing and discussing ideas is a way of life. I cannot say more than to thank my beloved family: Christopher Mathews, Marianne Mathews, Nathaniel Mathews, Benjamin Mathews, Rosemary Mathews, Sven Huseby, Barbara Ettinger, and Marya Huseby.

Finally, and always, my love and admiration to Kaia Huseby, and to Elias and Taddeo, who have made all of this worth doing.

Glossary of Institutions

CONAFOR *Comision Nacional Forestal:* National Forestry Commission.

EZLN *Ejercito Zapatista de Liberacion Nacional:* Zapatista National Liberation Army, which launched a rebellion against the national government in Chiapas in 1994.

FAPATUX *Fabricas Papeleras de Tuxtepec*: Tuxtepec Paper Factories.

IEE *Instituto Estatal de Ecología:* State Institute of Ecology.

INE *Instituto Nacional de Ecología*: National Institute of Ecology.

INI *Instituto Nacional Indigenista*: National Indigenous Institute. Responsible for development of indigenous communities, founded in 1949.

INIFAP *Instituto Nacional de Investigaciones Forestales Agrícolas y Pecuarias*: National Institute for Research on Forestry, Agriculture, and Livestock.

IRA *Instituto de Reforma Agraria:* Agrarian Reform Institute. Responsible for titling lands and ensuring the smooth administration of *ejidos* and *comunidades.*

LEGEEPA *Ley General de Equilibrio Ecológico y la Protección al Ambiente*: General Law of Environmental Protection and Ecological Equilibrium (1987). The first law to require an environmental impact analysis.

NGO Non-Governmental Organization.

NOM *Norma Oficial Mexicana*: Official Mexican Norm, the most enduring and authoritative of the regulations for controlling the environment.

ODRENASIJ *Organización para Defensa de los Recursos Naturales y Desarrollo Social de la Sierra de Juárez* (Organization for Defense of the Natural Resources and for the Social Development of the Sierra Juárez).

PAN *Partido Acción Nacional*: National Action Party. Center right political party that came to power in 2000, replacing the PRI.

PRD *Partido de la Revolución Democrática*: Party of the Democratic Revolution, a center left party.

PROCYMAF *Proyecto de Conservación y Manejo Forestal*: Conservation and Sustainable Forest Use Project.

PRI *Partido Revolucionario Institucional.* Institutional Revolutionary Party; remained in power from 1929 until 2000.

PROCEDE A land titling program run by the Ministry of Agricultural Reform, active in 2000–2001

PRODEFOR *Programa Nacional Forestal*: National Forest Program: Federal forestry subsidy program, active during 1994–2001

PROFEPA *Procuraduría Federal de Protección Ambiente*: Office of the Federal Environmental Prosecutor.

PRONARE *Programa Nacional de Reforestación*: National Reforestation Program.

SAG *Secretaría de Agricultura*: Ministry of Agriculture and Livestock.

SARH *Secretaría de Agricultura y Recursos Hidraulicos*: Ministry of Agriculture and Hydraulic Resources.

SEDESOL *Secretaría de Desarrollo Social*: Secretariat of Social Development.

SEMARNAP *Secretaría de Medio Ambiente y Recursos Naturales y Pesca*: Ministry of Environment Natural Resources and Fisheries (1992–2001).

SEMARNAT *Secretaría de Medio Ambiente y Recursos Naturales*: Ministry of Environment and Natural Resources (2001–present).

UCFAS *Unidad Comunal Forestal y Agropecuaria y de Servicios*: Community Forestry, Agriculture, and Services Unit—the organization responsible for managing Ixtlán's forests.

UNAM Mexican National Autonomous University. Until the 1980s, this was the preeminent educational institution in Mexico.

1

Introduction

Setting the Stage

In the film *Umberto D.* (Rizzoli et al. [1952] 2003), the title character, an elderly, retired civil servant, climbs onto a nearly empty tram. He is carrying a small suitcase and leads a small dog on a leash.

Conductor: *No no, col cane non si puó!*
Umberto D: *Prima delle otto si puó ...*
Conductor: *Lo insegna a me? Se é cacciatore sí, se non é cacciatore no.*
Umberto D: *Io posso dire che vado a caccia. Perche non potrei avere il fucile nella valigia?*
Conductor: *Va bene ... dove scende?*

Conductor: No, no, you can't travel with the dog!
Umberto D: But before eight one can.
Conductor: You're teaching me? If you are a hunter yes, if you're not a hunter no!
Umberto D: I can say that I am going hunting. Why couldn't I have a gun in my suitcase?
Conductor: All right. Where do you get off?[1]

Uncertain Authority

Around the world, the troubles of modernity seem to call for more knowledge, greater transparency, increased oversight by states, or increased inspection of states by active publics. It is often claimed that citizens should want to know more, perhaps in order to call governments and corporations to account, perhaps in order to make financial markets work better and avoid scandals and financial meltdowns. Global climate change, we are told, will be addressed by a transparent system of audit and accounting, which will make visible the stocks and flows of carbon from mines and forests into the atmosphere and oceans, hopefully preventing the worst impacts of climate change. International conservation organizations increasingly try to make biodiversity knowable to their audiences using brightly colored maps, which make visible where biodiversity is located, who or what is causing it to be eroded, and what, hopefully, might be done to address this predicament. Beyond the environmental field, efforts to produce transparent knowledge proceed unabated, from the calls of Transparency International to heed indices of governmental and corporate corruption, to efforts to monitor commodity chains that produce blood diamonds, to efforts to make high school teachers in Los Angeles accountable to quantitative assessments of their students' progress. Knowledge and transparency are key concerns across multiple cultures and problem areas, one of those things that you can never have too much of, even as you worry about the possibility of authoritarian states peering at the details of your personal life, or of oppressive bureaucracies that loose papers, demand taxes, and make your life complicated.

This book is about the effort to produce a regime of transparent knowledge in the forests of Mexico, and it is about how transparent knowledge was produced not by official declarations or scientific projects of mapping, but from the texture of encounters between officials and their clients, the foresters and indigenous people who manage and own the pine forests of Mexico. I will describe how the science of forestry arrived in Mexico in the late nineteenth century and how it gradually came to inform the lifeworlds of foresters, forestry officials, and indigenous people, and, more widely, how the political cultures of federal forestry institutions and their audiences affect how people believe or disbelieve in official knowledge about forests and about the state. This is a story about how transparency and other forms of knowledge are made; I will argue that when we talk of transparency or official knowledge, we too often assume that these are produced by officials in government offices or by scientists in laboratories. As I will show, in the case of Mexican forestry science, the apparently small scale and particular contexts of indigenous politics, logging in forests, and meetings between officials and indigenous leaders turn out to affect what we take to be very large categories: the credibility of the Mexican state, the stability of official knowledge about forests, the possibility of logging forests for timber. In other places and times, I will suggest, traveling theories are remade in local political performances; other regimes of transparency must deal with the power of publics to remake knowledge, to withhold belief in official beneficence and authority.

I have a confession to make. This is a book about forestry bureaucracies in Mexico written by an Anglo-American anthropologist, but I bring to bear on these social worlds a rather different sensibility. Growing up partly in Italy and partly in Mexico, I learned to see bureaucracies not as authoritative institutions that their clients obeyed, but as something quite different, as the sometimes dangerous, sometimes farcical and blundering instruments of the state. Everyone knew that a bureaucracy could be placated by a sufficiently persuasive performance, everyone had numerous stories to tell about their own encounters with bureaucracy, and everyone had better stories to tell than I did. In Italy I learned that if you could only present yourself as a peasant farmer, you might be able to secure tax exemptions and benefits from the state; I learned that the best way to approach an official was to secure his help in filling in forms or perhaps in avoiding forms and regulations entirely. Far from being an aberration or imperfection in the law, finding a bureaucrat to help or collude with you was the best possible way of negotiating with the state.

Years later, already in graduate school and studying Mexican forestry bureaucracies, I came across the wonderful films of Vittorio De Sica, the Italian neorealist film director. I came to realize that my sensibility of bureaucracy as an oppressive and malign fiction was imbued with an appreciation of the kinds of performances, collusions, complicities, and evasions that appear in many of De Sica's films. In the brief vignette I quote above, an old, unemployed official accompanied by his dog negotiates with a tram conductor in order to collude in producing a representation of a hunter leaving home early one morning with his dog. Somewhere the paper ticket that accompanies this story will leave a paper trail, and national statistics will refer to the number of hunters who use public transport. Documents here become potentially dangerous fictions, officials can be partially domesticated accomplices, and the state is far from being all knowing. This book recounts my travels within and encounters with the Mexican forestry bureaucracy and with indigenous forestry bureaucracies in the state of Oaxaca, but my point of departure was affected by the humorous or terrifying stories with which I grew up. Throughout this book, I describe bureaucracy as performance, as a public fiction, which can only be sustained by a skillful collaboration between apparently authoritative officials and their audiences, in a kind of public intimacy. I will argue that understanding forestry bureaucracies in this way radically transforms our understandings of modern states, of science, and of power. It is not that bureaucratic simplification and abstraction are the opposite of intimacy and collusion but rather that bureaucratic knowledge is always underpinned by collusion and intimacy, not just in Mexico but in other states and institutions. Official knowledge always silences other forms of knowledge, but this is not just a vice of bureaucrats: The literature on the sociology of knowledge teaches us that making shared public knowledge always involves silencing or suppressing alternative forms of knowledge. What is particular about bureaucratic knowledge-making is that it seeks simultaneously to perform official knowledge and knowledge of what kind of thing the state is. Officials silence opposition by claiming to speak for the state as thing and by claiming to translate generalized knowledge to local contexts, seeking to imprison their audiences in a slot of local knowledge.

Much anthropological study of conservation and development has assumed that these are powerful discursive forces that transform societies and environments around the world, through such projects as dams, road building, industrial agriculture, or the creation and policing of new parks. This is clearly a part of the story, but in this book I will argue that such accounts make conservation and development too powerful and fail to pay attention to the paradoxical authority and vulnerability, to the uncertain authority of conservation and development institutions and of modernist bureaucracies more widely.[2] Environmental anthropologists have given too much assent to the omnipotence and apparent omnipresence of conservation/development, perhaps framed as neoliberal conservation or neoliberal development, where it appears still more pervasive, more omnipotent, and still harder to oppose, both analytically and practically.[3] Often anthropologists frame their opposition to global forces as being a kind of speaking from the local, arguing always that local contexts are profoundly important and that globalizing projects are always reworked and transformed in local contexts. Valuable though this is, it imprisons the social sciences in a "local slot" that all too easily accepts the power of global generalizations and the institutions and actors who claim to speak for them. One way out of this conundrum is to pay close attention to the lives of the powerful, to look at how conservation officials, developers, or bureaucrats constantly juggle between local context and sweeping generalization, between the locality of their audiences and the global knowledge, general regulation or national policy they claim to speak for. This is what I call "uncertain authority": Officials may speak authoritatively, but they are haunted by a sense of vulnerability, as translating between the general and the local makes them vulnerable, worried about their lack of local knowledge. This book then is about Mexican forestry bureaucrats who juggle the tension between sweeping knowledge claims and mundane local concealments, between ambitious regulations and routine rule breaking. The power of these officials is different than we had thought bureaucratic power to be; it is a curious, halting, and vulnerable power, always made in performance, always subject to being undermined. This is an ethnographically observable, local, institutional power, which draws on the coercive and material power of a state that never reaches as far as it claims or would like to. State power rests on officials' ability to enact a distinction between the local and the global or the national, between a regulation and its specific local case, between the political and the technical. An attention to the detailed where, when, and how of bureaucratic lives and practices shows a more halting, less seamless, and more collaborative form of power, a power that seeks the assent of its audiences even as its performers doubt it. Seeing state power in this way calls upon us to rethink where, when, and how it might be fruitful to engage in remaking the state.

Following this insight, this book moves back and forth between the offices of the forest service in Mexico City and the regional capital in the state of Oaxaca, and the forests and the indigenous communities who largely control them. In chapter 2, I trace how the science of forestry first came to Mexico, how it was inscribed into national forestry laws and policies over the last hundred years, and I pay particular attention to the eminently material institutions and offices where particular officials were entrusted with the task of bringing forestry into forests. Forest policies and official forms of knowledge did not encounter a blank slate, and the new science of forestry encountered a landscape that had been partially transformed by histories of state-making and by past political economies. The details of colonial rule through indigenous municipalities and the struggles of indigenous people who engaged in warfare, trade, and state-making set the stage for the encounter between state science and popular understandings of forests, between forestry bureaucracies and indigenous municipalities. In chapter 3, I describe some of this stage setting, recounting how the landscapes and forests of the Sierra Juárez of Oaxaca were folded into economies of cochineal growing and mining in the eighteenth and nineteenth centuries, literally defining who would own forests in the twentieth century. The detailed histories of particular towns and forests turn out to matter a great deal for the credibility of forestry institutions in the present. I therefore focus particularly on the indigenous community of Ixtlán de Juárez, a small Zapotec town about a two-hour drive from the City of Oaxaca. Ixtlán was militarily powerful in the nineteenth century and came to be a leading forest community in the twentieth century. The details of ecology, landscape, and political history affect how indigenous forest communities like Ixtlán came to control areas of forest and how they brought to bear their ownership of forests in their encounters with the forest service and logging companies in the present. It is this experience of political action and of living and working with imperceptibly mobile forests and fields that indigenous people brought to their encounter with the new forest service in the 1930s.

In the wake of the Mexican Revolution (1911–1920), the expanding Mexican state brought the science of forestry and bureaucratic practices of paperwork to the City of Oaxaca and, haltingly, into the forests of the Sierra Juárez. Chapters 4 and 5 trace these moments of encounter, first when a relatively feeble forest service tried to directly control loggers through complex regulations and, from the 1950s onward, by subcontracting state authority to large parastatal companies that logged the pine forests of the Sierra Juárez and employed local people as forest workers and technicians. Working as employees of the logging companies taught indigenous people the theories and working practices of industrial forestry and gradually produced a popular movement that secured the cancellation of logging concessions in the mid-1980s. This marked a significant advance in the power of indigenous communities that owned forests, and it brings us to the present moment, when apparently authoritative state forestry institutions must deal with the mundane realities of limited resources, complex regulations, and intransigent local communities.

Paying attention to the daily work of officials allows us to see the curiously halting and hesitant power of officials and the power of their audiences. In chapter 6, I move from moments of encounter between officials and their indigenous audiences to the offices of the forest service in Mexico City, and then in chapter 7, I return to the lifeworlds of forestry officials and foresters in the City of Oaxaca. Crucially, I show that state power does not rely on knowledge alone but also on ignorance, and I argue that official knowledge and various forms of ignorance are coproduced in encounters between officials and their audiences. Local contexts and apparently local details turn out to matter a great deal for the content of official knowledge and for the legitimacy of government institutions. The power of forests and forest workers is further explored in chapter 8, where I describe how indigenous people in the community of Ixtlán are able to form alliances with government officials and with official knowledge, not because the state imposes legibility on them, but because a relatively powerful community is able to call community elites and government officials to account. Working in the forest becomes a political and an epistemic resource for loggers, technicians, and foresters—a way of reaching out and forming alliances with government officials, conservationists, and anthropologists. Finally, in a brief conclusion, I pose a set of questions rather than answers. I ask what it would mean for scholars of technology, bureaucracy, or the environment if we rethink knowledge as being always linked to ignorance and if we pay attention to the power of publics to affect official knowledge. If we see ignorance, collusion, and forms of nonknowledge as always having been embedded in making knowledge, how can we write about such ignorance and silences? What would this mean for our understandings of knowledge projects in the world, from economic projects of neoliberal reform, to projects to produce carbon markets or prevent the loss of biodiversity?

Indigenous Bureaucracy in Oaxaca

In August 2000, I sat on a bench outside the office of the mayor of the indigenous community of Ixtlán de Juárez, in the mountains of the Sierra Juárez of Oaxaca, in Southern Mexico. While I waited, a stream of people came and went; old men wearing plastic laminated straw hats and sandals, young men in baseball caps and sneakers, occasionally an older woman wearing the traditional black and white shawl (*rebozo*), in striking contrast to the mayor's elaborately dressed, made up, high-heeled secretary. I had come on a twofold mission: to ask permission to do research in the community of Ixtlán, and to ask the mayor why this small town of 2,000 people had become so successful in managing and protecting its extensive forests. I was not the first or last visitor on this mission; the indigenous communities of Mexico have become widely known for their sustainable forest management and have been promoted as a laboratory for community natural resource management for policymakers around the world (Bray et al. 2003; Bray, Merino-Pérez, and Barry 2005).[4] Like myself, researchers came to investigate how indigenous communities such as Ixtlán had managed to master the techniques of industrial forestry and to prevent the illegal logging that is so prevalent across Mexico and Latin America.

These two words, *indigenous* and *community*, have tremendous weight, a charisma that seems to be the opposite of impersonal state forestry bureaucracies. Around the world, state forestry bureaucracies are seen to be tainted by failure, because of corruption and lack of resources or because of conflicts between state and local people over the control of forests. I, like other researchers, wondered whether there was some special alchemy in these indigenous communities, some formula that might provide an antidote to the disenchantment and failure of state forestry. The word *indigenous* seems indeed to summon images of communal harmony and of living in balance with nature.[5] At first glance, an indigenous community appears to be the polar opposite of logging trucks, forest management plans, of form filling and of state forestry bureaucracies. In a similar way, the term "community" seems to be an opposite to markets, to the unceasing struggle for personal gain of modern capitalist society or to the depredations wrought by profit-seeking logging companies. How then had the community members of Ixtlán managed to square the circle, to balance modernity with indigenousness, capitalism with community? As I came to learn, in modern Mexico to be indigenous and to be a community member, has specific cultural and legal meanings that are policed by state bureaucracies and contested by popular movements and politicians. To be indigenous, then, is also to be modern and to be familiar with markets, capitalism, and individualism. Far from being a remote and distant, "natural" place, the forests and fields of Ixtlán had been affected by centuries of contact with the outside world, and its inhabitants were politically astute and skilled at negotiating with outsiders such as myself. Community and state institutions and forms of knowledge had in many ways been coproduced, and it was impossible to make sense of the forest community of Ixtlán without also studying the Mexican state. History literally mattered here: past events affected the structures of present-day forests; histories of struggles between communities and the state affected the content of official knowledge and the credibility of the Mexican state.

When I eventually entered the mayor's office, he was more than happy to talk to me. A friendly and cheerful man in his mid-40s, Graciano Torres[6] recounted to me how decades of destructive logging by the FAPATUX paper company[7] had taught community members to value their forests. He told me that community members had gradually taken control of every aspect of industrial forestry, from driving logging trucks and operating cranes, to managing a town saw mill, to marking, cutting, and replanting trees in the community forests. He described to me the process by which community members gradually learned to care for their forests:

> Before, exactly because we weren't culturally prepared, we thought that the paper company owned the forests, and we thought that when we cut trees and deliberately knocked down unmarked trees we were hurting the paper company, not ourselves . . . something which wasn't true, because we were harming our own

forest. But once forest management was passed to the people then we respected the [rules] even more, because people were oriented, the cutting areas were fixed. We have to respect the [cutting areas] and ask permission to cut trees, and we cut only marked trees. (interview notes, August 9, 2000)

Graciano's narrative of the incorporation of industrial forestry into local knowledge and practice contradicted images of indigenous communities as remote, traditional, and untouched, but in many ways it left unanswered as many questions as it answered. His account, of a transformation from a time when he and his fellow community members had not been "culturally prepared," strikingly agreed with the accounts of government foresters. Further, Graciano used the terms of scientific forestry in praising present-day community forestry, community loggers' respect for prescribed cutting areas, and for cutting only the trees marked by forestry technicians. How had this community assent to state forms of knowledge been produced?

The paper company to which Graciano referred, FAPATUX, was a parastatal logging company, a strange mélange of state bureaucracy and private company. Far from being opposed to the logging company, many of the older community members had been employed by it, and the community of Ixtlán had been one of the most reliable supporters of both the logging company and the Mexican state. The present-day status of Ixtlán as a model forest community had been produced not by distance from the state, but by the intimate and confused encounter between state and community. But what were the terms of this encounter? Did the state impose knowledge and practices on rural people? Or did people in Ixtlán and the Sierra Juárez appropriate and modify official knowledge in the interests of their own autonomy?

As I interviewed community elders, I began to realize that there had been a dramatic shift in people's understandings of their forests over the last generation. One such shift was a transformation in the way people understood fire. For elders, fire was a tool of agricultural management and could be controlled; for younger people, fire was uncontrollable and destructive, and a willingness to fight forest fires was one of the proud markers of community members' identities as protectors of the forest. How had this transformation come about? What combination of state or community coercion, official propaganda, and changing livelihood practices could have produced such a dramatic transformation in understandings of fire and of forests? What changing senses of self accompanied or contributed to this change? What was the role of state institutions in forming the proud identity of the people of Ixtlán as protectors of the forest? How had official knowledge and popular understandings of forests come to agree on the concepts and practices of industrial forestry so that present-day community members agree with government foresters that fighting forest fires is all-important? Community members had become adept not only in the state ideology of industrial forestry but in state practices; the forestry technicians employed by the community diligently filled in forms, detailing the locations, volumes, and species of timber cut. Did this mean that they were pinned down by an oppressive official gaze, which used state-defined knowledge to control rural society? Was the proud autonomy of the community no more than an illusion?

As soon as we see that the state was involved in the production of forest management in the remotest forests of the Sierra Juárez, it becomes necessary to question the very nature of state power and official knowledge. What *was* the state? Who were its representatives? What did they do, and how did local people respond to state interventions into their daily lives? How and when did official knowledge come to penetrate the consciousness of rural people? What were the terms of this engagement, and how much freedom did rural people have to modify or reject official knowledge of forests? Like other states around the world, the Mexican state is not a united structure; rather, it is a shifting group of loosely connected institutions that are unstable and often in conflict with one another. The state is not only a set of social structures, such as those optimistically represented by organizational charts; it is also the meanings attached to state power. This means that state-making requires continuous performance, a work that is always contested and never done.[8] What is the relationship between official performances and representations and popular understandings of forests? How do maps, organizational charts or officials' speeches become incorporated into or rejected from daily life in the Sierra Juárez? What is the relationship between official representations, routine bureaucratic practices, and people's identities and political engagements?

This book tries to answer some of these questions through an investigation of how Mexican political culture has affected socially accepted knowledge about what forests are and what the state is. I will argue that forestry officials have continuously tried to perform the state as the kind of thing that is beneficent, knowledgeable, and unified, and that these performances define the contours of the political, of what can and cannot be said. I will suggest that the arrow of influence is not one way and that the texture of local contexts of state-making powerfully affects what officials come to know through their daily paperwork practices. I will argue that officials and their audiences share understandings of the state as a dangerous illusion, and that public framings of official performances of knowledge also affect official efforts to perform the state as unified, knowing, and beneficent. I will describe the long-term coproduction of community and state power, combining an investigation of national forestry institutions and offices in Mexico City and Oaxaca, a detailed analysis of practices of state-making in the forests of the state of Oaxaca, and an ethnography of the Zapotec indigenous community of Ixtlán de Juárez, currently one of the leading forest communities in Mexico. I will describe how industrial forestry has come to be part of community identity, as manifested in fire fighting, forest management, and road building, and in local conceptions of nature and culture. This was not the imposition of an authoritative forestry bureaucracy on a more or less passive society. On the contrary, I will show that rather than being the product of ideological domination or direct coercion, the transformation in indigenous people's understandings of forests was the product of community political power and autonomy. Powerful forest communities have been able to form alliances with forestry officials to coproduce socially accepted knowledge about forests. Far from being powerful and authoritative, forestry bureaucrats in Mexico have experienced frequent institutional reorganizations and are haunted by a sense of doubt, of not knowing.

State-Making and the Production of Knowledge and Ignorance

A generation of research on the state has shown that far from being unitary and monolithic, the institutions that are supposed to implement technical knowledge and development[9] are fragmented, hierarchical, and unstable. Similarly, over the last twenty years, research in the anthropology and sociology of knowledge has revealed that scientific and technical expertise is an often fragile achievement, produced by building networks of alliances between scientific data, material objects, and researchers in different laboratories. In the case of Mexican forestry, pine seedlings in forests in Finland and France are linked to FAO forestry experts, government officials in Mexico City, forestry regulations and management plans, local-level forest police, and logging practices in the Sierra Juárez.[10] What is striking about the network of connections that supports the science of forest ecology is not how powerful and stable it is, but how *unstable* and fragile it is. At any place in the network, it appears easy to conceal, avoid, obfuscate, or hide. As the stakes of concealment rise, it appears more and more likely that deliberate concealment or mistranslation may fatally weaken the network. How then can government officials, environmental activists, and loggers in the Sierra Juárez come to share a similar understanding of what forests are and how to manage them? If the stakes in mistranslation or concealment are high, it seems likely that state-sponsored environmental discourses will be evaded and will have little success in transforming popular identities and practices.

States around the world have based their power in part on claims to knowledge, making the stakes of knowledge very high and making official ignorance correspondingly valuable. For every official attempt to control rural people through a tree-cutting regulation or a map, there is a corresponding incentive for rural people to avoid, conceal, or escape, whether through the classic weapons of the weak (Scott 1985) or through more directly forbidden behavior, such as illegal logging or agricultural burning. In many cases, these evasions of official discourse take place within the very state apparatus that is supposed to enforce it. In the case of Mexican forestry, as I will show, officials bypass or selectively enforce forestry regulations that they believe to be impractical, controversial, or misconceived. In this book, I will argue that the hierarchical power structures of the Mexican forest service and the menace of state power have caused profound official ignorance, not only of people's motives and intentions, but of their most basic daily practices. Paradoxically, official knowledge is produced not by the menacing power of the official gaze, as manifested in the census and the cadastral map, but by the more or less willing assent of rural people in the forms of knowledge and politics. It is not the case that evasion, collusion, or foot-dragging is the opposite of state power. On the contrary, performances of authoritative simplifications are underpinned by collusion, silence, and evasion. Where power relations between state and rural people are not too uneven, where there is sufficient autonomy, such as when forest communities are well organized and can assert themselves against inadequate regulations and official interference, then official and popular knowledge may be shared, forming an epistemic community of shared knowledge and action.[11] Knowledge then is underpinned by an alliance, by a shared understanding of the world. Such understandings can take the form of "boundary objects," shared forms of knowledge that allow autonomy and differences between allies or collaborators (Star and Griesemer 1989). In Mexico, one such boundary object is the understanding of what forests are and how they should be managed, which is shared by the forest service[12] and some rural communities. On the contrary, if people feel themselves too much disadvantaged by official regulations and conceptual definitions, they are likely to pay regulations only lip service and ignore them in their daily lives. An example of this is the widespread practice of agricultural and pastoral burning, which is ubiquitous in Mexico, although officially forbidden.

Scientific and technical expertise are often thought of as the willing servants of the state, assisting the advance of state power through building dams and irrigation systems, promoting pesticides or modern medicine, displacing local knowledge with frequently disastrous results (Scott 1998; Mitchell 2002). In this account, modern science and the modern state advance like a steamroller, crushing or coopting local opposition beneath the juggernaut of progress, defining out of existence the expertise of rural and indigenous people who are characterized as "backward" or "ignorant." In a similar vein, critics of development have pointed to a development apparatus that creates underdeveloped subjects for development; these subjects are then crushed or displaced by the advance of an apolitical and purportedly neutral development machine (Ferguson 1994; Escobar 1991; Goldman 2001). These accounts are helpful insofar as they make official knowledge a central concern and reveal it to be a richly cultural practice, denying its claims to generality, impartiality, and distance from the "local contexts" in which anthropologists and others live and work. However, this can only be a beginning: Ethnographers of official knowledge have too easily accepted modernist bureaucracies on their own terms, as more or less unitary institutions that gather knowledge, classify it, render it technical, and then act on nature and society in the name of that knowledge. This critical anthropology of development has inverted official rhetoric of knowledge by commenting with horror on official ignorance of politics (Ferguson 1994; Arce and Long 1993; Hobart 1993; Van Ufford 1993) or local ethnographic details (Li 2006:3). Such criticisms too easily accept modernist bureaucracies' rhetoric of general or abstract knowledge, even as they criticize them for failing to live up to their proclaimed projects. A more radical critique would focus in detail on the daily practices of bureaucrats who perform abstract and general knowledge *against* the audiences who they seek to make local, to look at how making knowledge and ignorance are partially intentional practices, and to take seriously science studies' scholars' insight that making knowledge *always* requires the silencing, ignoring, or suppressing of alternative kinds of knowledge (I will discuss this in more detail below).

Analyses of official knowledge as a power-laden discourse fundamentally misconceive the relationship between power and knowledge because they pay no attention to materiality, practice, and resistance and too easily assume that official discourses are uniformly internalized by government officials. Official discourses, in these accounts, are like an invisible fluid that permeates all officials and often their audiences. This pays too little attention to the internal fissures and tensions *within* the state and to the material daily practices of bureaucrats who shuffle papers, annotate reports, and sign permissions. An official discourse may mean very different things to the minister of environment who pronounces an oration, to the field-level forester in Oaxaca who pretends to enforce a forestry regulation while actually ignoring it, and to rural people in the Sierra Juárez of Oaxaca who repeat official language to visiting functionaries. Overemphasizing the power of official discourse pays too little attention to the daily practice of politics within and outside state institutions, to the informal networks of patronage by which officials, politicians, and ordinary people seek to appropriate or modify the power of the state. This work of politics and career building is not just a failure of modernity or a result of corruption. Rather, this is the way that people in state institutions in Mexico (and in many other countries) make sense of the tension between official knowledge and their daily work lives; these evasions make sense on their own terms and are widespread in all modernist bureaucracies.[13] This leads me to argue that understandings of knowledge as a uniform discourse are less useful than a formulation of knowledge as practice and performance. Over the last hundred years, Mexican forestry officials have struggled to perform the state as a certain king of thing: as a unified, beneficent, and knowing institution that can know what happens in distant forests and reaches uniformly into the furthest reaches of the forest. Focusing in this way on knowledge as performance draws attention to the power of the audience to believe or not to believe, to the distance between the performers' on- and offstage assertions, to the skill required to produce an effective performance, and to the political costs of failure.

A rather different approach to official knowledge is taken by James Scott, who draws attention to the aesthetic beliefs and desires of officials within modern states who seek to remake society and nature in ways that make sense to them (Scott 1998). Scott shows how authoritarian states have sought to impose simplified and officially legible landscapes on prostrate civil societies, describing the catastrophic failures of forced villagization in Tanzania and collectivization in Soviet Russia (Scott 1998:4–5,193–260). For Scott, officials have interests and aesthetics of their own (Scott 1998:18); they constantly straddle legibility and illegibility as only the kinds of simplifications administrators wish to know are recorded (Scott 1998:11), whereas official practices are often sustained by a "dark twin" of illegal or informal practices in which officials may collude (Scott 1998:331). Suggestive as this is, Scott largely takes for granted the ability of states to imbue officials with the desire to impose projects of legibility and visibility, while we do not see how they go about concealing evidence of failure from their superiors or themselves (although this is strongly suggested by their willful persistence in failed policies). Scott's own earlier work on resistance and "weapons of the weak" (Scott 1985, 1990) sits awkwardly with the unity of official projects of legibility in "Seeing Like a State". I suggest that resistance and foot-dragging are not necessarily the opposite of official projects of legibility, but rather that they are the ground upon which performances of the state and of official knowledge take place. As recent ethnographies show us, officials may ignore government ideologies and projects (Li 1999); they may carry out rituals of assent even as they undermine regulations by their daily actions, or they may collude with the subjects of rule from sympathy, for personal benefit, or from political necessity (Herzfeld 2005:375). Paying close attention to these mundane practices of collusion and evasion radically transforms our understanding of the location and texture of official knowledge-making and even of the project of legibility itself. Rather than an official knowledge that arises from the imposition of legibility on officials, society, and nature, as Scott describes, I will show how official knowledge is the relatively fragile product of negotiations between officials and their audiences in meeting halls and offices. The detailed descriptions of encounters in Mexican offices and forests allow me to make a more general claim that officials in other places and at other times may decide to ignore projects of legibility. Transparent knowledge is a dream of modern state institutions, and officials in other places and contexts may deal with their political weakness by seeking to entangle powerful allies in official knowledge claims and by concealing their own activities from their superiors.

Seeing and Being Seen: State Formation and Identity

But what is the relationship between these routine practices of bureaucratic power and the diffusion of state ideologies into society?

Foucaultian conceptions of power/knowledge, as a set of rules about the production and circulation of official knowledge (Foucault 1991), do not do justice to the internal conflicts *within* the state. How is knowledge translated from one level to another, and what transformations, concealments, and betrayals does it undergo? In chapters 6 and 7, I pay particular attention to the culture of concealment and accommodation within the forest service and to the ways that government officials and their clients subvert or ignore forestry regulations. This leads me to conclude that the enduring consistency of official environmental discourses is the product not of stable and enduring bureaucracies but of the weakness and instability of state institutions. High-level forestry officials retain control of the symbolic capital of regulations and official environmental discourse even as they wrestle with their material inability to enforce these regulations and their doubts as to whether their subordinates are obeying their commands.

A better guide to the inculcation of new identities lies in paying close attention to the density and texture of encounters between officials and their clients and in comparing these with the daily practices where these clients in turn engage in daily life. How often do forestry officials meet with rural people? Can they really enforce regulations or do they merely pronounce them and then proceed to ignore them, as in the case of regulations forbidding agricultural fire use? In contrast, what are the daily practices of rural people? Do people make a living in ways that are officially forbidden but necessary to daily life? Agropastoral fires are necessary to agricultural practices over much of Mexico and take place within a sphere that is deliberately concealed from the attention of the state. It is not likely that fire users have internalized state understandings of fire through their encounters with officials; indeed, their encounter with forests is as important as is their encounter with the state. The environmental identities of people in Ixtlán are produced not only by mainly state-produced environmental theories but also by the logging practices in which they engage and through their encounter with the stubborn resistance of the natural world—the trees, forest fires, and logging roads with which community loggers must engage. In this context, nature is an actor that in turn affects the identities of human actors. This is suggestive of the power of practices of bureaucratic paperwork, which encounter material or conceptual resistances and may offer similar possibilities for distancing bureaucrats' identities from official projects of knowledge or control.

Documents, Material Visions, and the Cloud of Lies

I have long been haunted by an image of how documents that purport to reveal end up confusing and concealing (see figure 1.1). This picture, a 1940 lithograph by the eminent Mexican artist Jose Chávez Morado, is an image of concealed danger, transformation, loneliness, and isolation. The figure in the foreground is a worker in overalls (perhaps he is an industrial worker?); a storm of newspapers wraps his head and has concealed an abyss into which he is about to step. Elsewhere across a bare and empty plain, other figures wander alone, struggling blindly with their own storms of newspapers. All are isolated from each other, none can communicate, and all are endangered. The large Frankenstein-like hands that reach blindly forward suggest another kind of danger, in a presentiment of monstrous transformation. Another kind of metamorphosis is artfully alluded to: Newspapers that might transform human beings into documents are a visual echo of classic paintings and sculptures of Daphne turning into a tree when pursued by Apollo.[14]

Morado was warning against what he saw as the lies in official newspapers and against state efforts to delude labor unions in the 1940s. However, during the course of this book, I will suggest that this image of dangerous public illusions has a continuing contemporary significance, and that it illustrates an enduring cultural framing of public knowledge that officials and their audiences bring to bear in performances of public knowledge and of the state. This imaginary of documents that conceal hidden danger vividly illustrates not only how publics view the state but how officials themselves view the documents which they handle.[15] Forestry officials in Mexico must act as if they believed the content of the documents that are their daily companions, but they are haunted by the sense that these documents are lies that may conceal a hidden pitfall that will cost them their jobs. Official efforts to make Mexican forests transparent and legible have always relied on documentary practices, and vision and supervision have always relied on material papers.

By focusing on the materiality of documents and forms, and on officials' complex calculations of how to deal with such regulations, I show how precisely those documents and forms that seek to produce transparency produce their opposites, concealment, and official ignorance. Official practices of transparency and visual supervision can become the storm of papers, in which neither officials nor their audience believe. In addition to its specifically Mexican associations, I suggest that *Cloud of Lies* can also be used to rethink the ways that vision is often used in

Figure 1.1
The Cloud of Lies: *Nube de Mentiras*, lithograph by Jose Chavez Morado, Mexico City, 1940. *Source:* Reproduction courtesy of the Philadelphia Museum of Art, gift of R. Sturgis and Marion B. Ingersol, 1943.

political theory. This image suggests that we can think of vision, seeing, and knowledge-making as profoundly material practices. Seeing vision as material in this way radically undermines the metaphor of vision as unmediated, direct perception, so familiar from discussions of politics and the state. The documents that blind and confuse, which conceal the abyss, are made of papers that are physical, tactile, and real. When I look at this picture, I think of my own past efforts to capture papers blown by the wind, newspapers that crumpled and escaped my grasp, and essay drafts that I had to chase across a parking lot or an office.

Material Visions, Official Knowledge, and Ignorance

Since its inception in early modern Germany (Scott 1998; Rajan 2006; Nelson 2005), Italy (Appuhn 2000), and in the British, French, and Dutch colonial empires (Grove 1995), forestry and conservation have been quintessentially state activities, and states have sought to assert their authority over forests through rhetorics of legibility and transparent knowledge. Controlling nature for economic, strategic, and environmental reasons has been part of the constitution of modern states, and performing the control and legibility of nature has been one way in which rulers have tried to establish the stability and reasonableness of rule. Forests, often conceived of as being one of the wildest aspects of nature, are typically remote from the rulers and officials who seek to control them, and forest-dwelling people have often been seen as problematic, ethnically other, and dangerous. More recent efforts to control and regulate nature, from biodiversity mapping to designing carbon markets that might prevent climate change, can learn something from the history of forestry and of Mexican forests. More generally, the history of forestry can offer lessons for those who are interested in the ways that producing public knowledge can legitimize institutions, from efforts to reform financial markets to efforts to reform states in the name of neoliberal economics.

Official knowledge of forests and of people has often been assimilated to the metaphor of vision and to associated practices of supervision and control. In much of political theory, vision and associated terms ("legibility," "transparency") are used somewhat unreflectively as metaphors for unmediated direct perception, for a direct knowledge of what is going on, a kind of knowledge that does not require the observer to interact with the people or places being observed. James Scott, for example, uses "seeing" in the title of his wonderful book, *Seeing Like a State,* along with his evocative coinage of "legibility," in order to describe the efforts of modernist bureaucracies to make landscapes legible to visual inspection, taxation, and control. Another use of vision as a metaphor for power comes from Michel Foucault's famous discussion of panopticism (Foucault 1979), where the subjects of rule internalize the possibility of visual inspection even when it is no longer occurring. Here, too, visualization and inspection are metaphors for a kind of unmediated direct knowledge and control. This kind of power is perhaps the dream of the powerful: to know others without being known to them, perhaps through what Haraway calls the god trick of a disembodied knowledge that has the quality of a view from nowhere (Haraway 1991), perhaps through Haroun al Rashid's mythical desire to walk incognito through the streets of Baghdad. All too often, political theorists seem to confuse this desire and the associated rhetorical claim of transparent knowledge, with its effective reality, official knowledge of a legible and transparent society. Vision as direct knowledge is a troubled metaphor because it erases the materiality of seeing: This erasure in turn makes it possible to imagine seeing without being seen. One way to restore the materiality and interactivity of seeing and knowing is to question seeing, to make visible the material objects (documents, forms) and social relations that make seeing possible, or to use metaphors of vision as touching, as when we are blinded by documents that purport to reveal.

It is all the more significant and troubling that vision is so unambiguously associated with knowledge, perception, and control within political theory because in other fields, seeing is seen as profoundly complex and problematic. Within science and technology studies, many scholars have pointed out that audiences have to be taught how to see (Daston and Galison 1992; Dumit 2003) through performances of public reason, expert authority, and the use of material images (Jasanoff 1998; Shapin and Schaffer 1985:22–77). In recent work, Haraway talks of visual prostheses, and of "optic-haptic" vision, seeing as touching by "fingery eyes" (Haraway 2008:250). The history of western optics, with its emphasis on ray theories, where the independent observer's eye captures rays of light emitted by the object that is being seen, are more confusing than helpful here. Karen Barad calls for a diffractive and intra-active kind of seeing, which draws on wave theories of light and a sophisticated discussion of quantum mechanics and complementarity (Barad 2007). These rethinkings of vision compel a rethinking of political metaphors of visuality as power. Officials who seek to make landscapes inspectable and legible must engage their human and nonhuman interlocutors: Unmediated vision is a political fiction or a description of the kind of knowing that emerges from hard political and epistemic work. Such kinds of socially accepted knowledge require forms of assent from their audiences. Such assent can take the form of collaboration, collusion, dissimulation, or doubt.

The case of repeatedly foiled efforts to produce transparency in Mexican forests is relevant not only to Mexico, but to our understandings of the relationship between bureaucratic authority, institutional power, and knowledge, and to the power of publics in the making of knowledge. Efforts to produce transparent knowledge of nature or, more widely, knowledge of "the way things are," are currently widespread across a variety of fields. In biodiversity conservation circles, multiple efforts seek to use satellite images and remote sensing to drive ecoregional planning (Brosius 2006). Similarly, numerous scientists and policymakers around the world are engaged in an effort to make forests legible and visualizable to world carbon markets (Bumpus and Liverman 2008). Should such a project of seeing succeed, buyers of carbon credits in London or New York would buy and believe in carbon futures in order to pay distant farmers for the carbon sequestered in their trees and soils, secure in the belief that this carbon capture was visually guaranteed through quasi real-time satellite surveillance. Recent events in global financial markets demonstrate that here, too, transparency as a metaphor for unmediated knowledge of reality is a key term. Financial meltdown is blamed on the "lack of transparency" in new financial products, where buyers did not know the risks that they were buying, and unreliable intermediaries pocketed huge fortunes. Many critics of recent financial scandals have suggested that the best means of preventing further economic disasters is through regulations that will make markets transparent and will allow publics (often framed as investors) to know that bank balance sheets do not conceal hidden toxic assets. Here, too, knowledge of reality is framed as transparent vision, and here, too, a complex web of financial operations, calculations, and regulations will, it is hoped, produce credible knowledge of reality, which will come to be seen as having a kind of visual certainty. In all such projects of visualization, intermediate material instruments, documents, and people disappear: The moment of knowing and perceiving effaces the scaffolding that made vision possible.

My goal here is not to purge politics of the metaphor of vision and transparency but to describe more clearly how vision describes the kind of knowing that happens after much political and epistemic work, when the material and political supports of knowing disappear from the consciousness of the knower. For example, before I studied forestry, I saw forests as a more or less undifferentiated green wall; the differences between trees were visible and yet hard to remember, hard to discern. After (some) training in systematics and taxonomy, from reading books and walking around with a teacher, I began to see forests differently. At first I had to stop and check every tree, painstakingly looking at leaf characteristics, bark, or flowers, but eventually I could look at a tree from a distance, somehow putting together bark, leaves, color, and a host of other details so as to see the tree as a red maple, an olive tree, or a redwood. At this moment, all the hard work of reading, walking, and talking disappeared; seeing then could happen when material practices and histories became effaced in a moment of recognition. Such practices of seeing are skilled and never definitive, the world does not necessarily sort into easily distinguished species, as in New England forests where oak species interbreed, producing a "hybrid swarm" that undermines the value of the species concept. When we all agree that knowledge is good and real, it comes to seem transparent, but this transparency always relies on such practices as walking and looking, on practices of paper work or audit, on performance, and representation. Official efforts to describe official knowledge as vision are rhetorical claims, efforts to assert unmediated knowledge, but they are more a desire than a reality.

Making Things Technical, Making Things Political

For scholars of development, making things political or technical is a key moment in the assertion of rule, but this making of the technical is too often assumed to be successful in hermetically closing off the political from the technical (Ferguson 1994; Mitchell 2002). This is at odds with much of the literature on science and technology: For science and technology studies (STS) scholars, the boundary between the political and the technical is continually contested and remade (Gieryn 1995). In Shapin and Shaeffer's *Leviathan and the Airpump* (Shapin and Schaffer 1985), the technical must be performed and witnessed and is always defined *against* the political. The authors describe Robert Boyle's role in defining scientific expertise through practices of performance and witnessing, where scientific knowledge was defined as knowledge produced before qualified witnesses. Stephen Hilgartner follows this dramaturgical metaphor and argues that expert and scientific advice are always a kind of public drama (Hilgartner 2000), staged by scientists and officials in an effort to command the assent of the audience. What Hilgartner and other science studies scholars make clear (e.g., Wynne 2005:85) is that performances of expert knowledge are *always* public even or perhaps especially when they proclaim themselves most distant from politics and witnessing audiences. Such performances of expertise do the political work of defining a narrow audience of expert witnesses and of defining what subjects are open to political discussion. These dramatic performances of expertise seek to define the role of nonexpert publics as passive witnesses, who nevertheless must still assent to expertise, performance, and public reason.

Public performances of scientific knowledge define the contours of the political by making and remaking the boundary between science and politics. For science and technology studies scholars then, defining the technical always involves defining the political, and the technical and the political are always coproduced (Jasanoff 2004, 2004; Latour 1993). Each redefinition of the technical redefines expertise, the role of audiences, and forms of witnessing; it also redefines how and where political debates about justice can take place.

This conversation can be brought to bear on the critiques of technocratic knowledge-making within anthropology. James Ferguson (Ferguson 1994) and his interlocutors (e.g., Escobar 1995) have argued that development experts seek to define development as an apolitical intervention. For them, the scandal of knowledge is that political decisions are made in distant smoke-filled rooms or government planners' offices, where supposedly impartial technocratic knowledge improperly conceals something entirely different. In such accounts, the public has been effectively excluded from making knowledge, and technocratic expertise has effectively done its work so that corrupt elites or indifferent officials reap economic rewards, succeed in entrenching state domination, or disregard pressing political claims and movements. These accounts of anti-politics miss the public nature of anti-political performances that seek to define who participates in knowledge-making and on what terms. Development experts and government officials' performances of technocratic or planning knowledge seek to coax the public to become a more or less passive witness to distant and already completed performances of expert knowledge. Even the most apparently anti-political of knowledge claims seek to make claims on the public, and such performances are unstable and potentially fragile. Anti-political knowledge is not a seamless discourse nor even a unified project of producing legibility, but rather a potentially fragile performance that seeks to make both the technical and the political. Recent work by Tania Li, which shows the fragility of performances of knowledge by development experts in Indonesia, is a powerful pointer (Li 2006). This suggests collusion and complicity between experts and their publics and draws attention not to the hegemony of official knowledge, but to the reversals, confusions, and moments of upsetting, when officials scratch their heads and change their stories. Here, audiences become powerful actors who can accede or refuse assent to these dramas.

What then is the role of audiences? What resources do the audiences for public knowledge-making have for resisting, affecting, or reinterpreting knowledge performances? This turns our gaze to political culture, to enduring public framings of the state, of expertise, performance, and of what expertise should look like. Sheila Jasanoff calls these cultures of public knowledge-making *civic epistemologies*, drawing attention to the ways that publics are always involved in the coproduction of politics and knowledge (Jasanoff 2005). The term "civic" might be problematic for anthropologists, suggesting a normative concept of the proper forms of citizenship, and perhaps of a problematic separation of state from civil society. However, I suggest that we can take from this not a normative claim that civic engagement is proper, but rather a prediction that engagements between states and other actors in fact take place in a variety of places and in ways that do not necessarily appear very civic. As we shall see, Mexican publics are skeptical and unwilling to openly voice their criticisms of official knowledge-making, but this does not mean that they believe official pronouncements. Civic engagement in this case takes the form of public deference and a large measure of disbelief in official performances of knowledge. Mexican officials and their audiences see the state as dangerous and official knowledge as a mask, an illusion that conceals possible personal dangers. This framing of official knowledge as performance and illusion affects not only how scientists seek to perform knowledge before publics, but efforts by politicians who seek to perform the state as knowledgeable, beneficent, and unified.

In taking seriously the state as *thing* and in comparing officials' public performances of official knowledge to the knowledge-making practices of natural scientists, I go in a different direction from much recent anthropology of science. In such works as *When Species Meet* (Haraway 2008), *Alien Ocean* (Helmreich 2008), or *Dolly Mixtures* (Franklin 2007), there is little or no mention of the state. The authors are more concerned with how new kinds of science change what it means to be human and with the power of speculative futures to create new forms of capitalism. I take seriously these scholars' concerns with materiality and knowledge-making, but I turn my gaze on the materiality and performance of the state as an object of knowledge, as a *thing*, an empirically traceable set of institutions, documentary practices, and bureaucratic lifeworlds. A theme that runs constantly through this book is my effort to keep track of what the state was at each moment, how many forestry officials, how many technicians, where they lived, and how much practical power they had. Here, theory informs method: We cannot take the power of the state for granted, and we have to weigh it carefully at various moments, from the fragile moment when forestry science arrived in Mexico City to the present moment when forestry bureaucracies are widely spread across the Mexican landscape.

Writing Resistance Into History: Nature, Culture, and Human Agency

As a study in environmental state building, this book traces the construction of nature/culture boundaries by the Mexican state over the last hundred years and shows how particular constructions of nature have been deployed to bolster the legitimacy of the state. Collective representations of nature have been continually remade in response not only to state projects, but to local practices of meeting and working with other living and nonliving things. Pine trees, roads, documents, and chainsaws are all stubbornly resistant material things and active participants in human projects. Loggers, farmers, foresters, and road builders come to know themselves in encounters with corn plants, marking hammers, documents, and wet roads. People bring these kinds of knowledge of self and the world to their encounters with the state. This means that officials' efforts to perform the state as a stable, knowing, and powerful agent encounter an active, knowing, and often skeptical audience. This audience brings to bear its own cognitive and epistemic resources in accepting, undermining, remaking, or evading official projects of making knowledge. Officials seek to entangle their publics, to produce assent to official knowledge claims and allegiance to official projects, drawing on enduring framings of how knowledge is supposed to be produced. Representations of official beneficence, or of environmental degradation, are not necessarily accepted by the subjects of state control, as in the Sierra Juárez of Oaxaca, where indigenous communities have imposed their own counterhegemonic history of environmental change on the forest service.

Unifying nature and human agency within a single frame of analysis poses problems of knowledge and method. How can we integrate the different forms of knowledge of the natural and social sciences without prejudging the primacy of one over the other and without a naïve positivism that asserts true or correct knowledge about the natural world? In contrast, most social science accounts of society implicitly neglect nature as an actor, either ignoring it or depicting it as a social construction (Mitchell 2002:19–53; Latour 1993). This problem is of central interest to any environmental anthropology or environmental history, to any study of science and society. In a real sense, a study of social and environmental change without an active and intransigent nature is a drama stripped of its principal actors. How can we make sense of the lives of people who struggle to make their livings from forests and fields if we do not pay attention to the material/ideological conditions of that struggle, if we end up saying that what really matters is their relationship with the state, with each other, but not with their fields and trees? Theories of knowledge as performance and practice, rather than as a representation of the world, provide a working method, if not a complete solution. Such theories also help us think of the remaking of humanness and highlight how making natures produces new subjectivities, refusing to make state-imposed identities the most important or only story to tell.[16] In a real sense, human agency with regard to the environment can only be described by giving nature a corresponding agency of its own—a kind of unruly obstinacy, which sometimes frustrates human projects and interpretations. The unruly obstinacy and liveliness of nature is a resource for people who go about making knowledge of who they are, what the world is, and what the state is.

In this book, I have addressed this dilemma by drawing on an eclectic variety of methods from the natural and social sciences in the hopes of destabilizing the power of any particular method (e.g., Rocheleau 1995) and of revealing the limits of each form of knowledge. I have made a pragmatic use of different forms of knowledge while attempting to remain aware of the limitations, theoretical assumptions, and *resistances* that each method encounters. Using multiple forms of knowledge highlights their associated theories and methods and destabilizes the dominance of any particular discipline. I do not use these multiple methods with the aim of a kind of triangulation that will produce a composite representation that will be closer to the "real" world. Rather, the multiplicity of methods and theories reveals the limits of knowing and the multiple resistances that knowing encounters. I have long been troubled by a tendency in the social sciences to seek to explain too much: ethnographically thin writing in which human actors emerge in order to explicate a theoretical point, only to disappear once again. This is very contrary to the ways in which many natural scientists locate what is not explained. In my former disciplinary training as a physicist and then in forest ecology, I learned to display graphs and charts, in which neat lines were not undermined by the presence of data points far from the line. Natural scientists, within limits, are comfortable with the presence of unexplained variation on graphs of results. On the contrary, results that fit a predicted line too well would be greeted with suspicion and doubt. For me resistance to knowledge is a sign of practical knowledge itself, of the liveliness of the world, of the unpredictability of people, pine trees, and fires. This attention to resistance is a thread that runs through the book, from the histories of shifting pine forests and fields, to the resistances that officials encounter when they try to perform authoritative official knowledge in the face of public skepticism.

By paying detailed attention to the constraints of ecology and climate (especially in chapters 3, 4, and 5) and by carrying out a small fire ecology study, which I describe in detail in chapter 3, I have paid close attention to the resistances that the natural world offers to both my own analysis and to the projects of foresters and farmers in the Sierra Juárez. This emphasis on practice and *resistance* to interpretation is an important clue to a kind of writing that gives agency to the subjects of ethnography and to the natural world.

In my early 20s, while traveling in Peru as a young and feckless backpacker, I experienced a small earthquake while staying on the top floor of a small hotel. My friends and I felt the building move, contrary to our experience of nature, a strange, unsettling, and yet exhilarating sensation. Perhaps because we didn't know how to sense this experience as danger (fortunately no one was hurt), we fell off our beds laughing. The world was alive—it was unexpected and unforeseeable. In my training as a natural scientist and later as a social scientist, I have tried to follow this insight—that the world is lively. It is this impulse that seems to me to drive the curiosity of natural and social scientists alike. In addition to providing the material for explication and analysis, I suggest that looking for ethnographic surprise can reveal the excess of human beings and of nature. Too much writing about science, nature, and people makes implicit claims to an impossible degree of knowledge and erases precisely what is interesting about people and things, their lively and agentive quality. Discourses and political economic structures may limit human agency most powerfully when we let them discipline our writing, eliminating confusing and lively people. It is the stubborn resistance, autonomy, and unpredictability of real human subjects that makes an ethnography or a history convincing and that makes life interesting. In a sense, while ethnographers can explain, as when we describe what people said to us and the kinds of reasons that they gave for what they did, we must avoid *explaining away*, where the ethnography contains only what fits our theories and where the people described are shadowy puppets who emerge on stage to illustrate a theoretical point, only to be skillfully removed once their work is done. I suggest that surprise and excess are the markers of an ethnographic practice that takes seriously the agency of people and nature. In this book, I have tried to write such excess into the text. There is more going on in Mexican forests than I can explain: I have tried to write some of that more into this book.

Methods

My journeys through the Mexican forest service were initially made easier by my own ambiguous status as both a forester and an anthropologist, and I conducted in-depth interviews with forestry officials in Mexico City and Oaxaca as a complement to participant observation and interviews in forest communities in the Sierra Juárez (see appendix 1 for a list of interviews). Although most of my work was carried out in 1998, 2000–2001, and 2003, return trips in the summers of 2008 and 2009 allowed me to trace the significant continuities between the forest service as it was when I first encountered it and how it looks now, after political reforms, budgetary crises, and new fashions in administration and conservation practice. A number of senior Mexican forestry officials had completed PhDs at the Yale School of Forestry, where I was studying, and it was they who I initially approached. Often I was cast in the role of a junior colleague, a guest who would accept their hospitality and listen to their narratives about life in Mexico, about life as a forester, about life as a functionary. This was a new experience for me: At graduate school, I was expected to be enthusiastic and attentive, indicating my interest by producing ideas for my professors. In my initial meetings with the senior foresters, in forest service offices in Mexico City, and at the University of Chapingo, I learned to play the new part of a deferential and relatively silent junior. In some ways this was ideal; they were happy to talk to me as long as I was prepared to listen and I could take copious notes. These former Yale forestry students were by then working as senior officials and in universities, forming part of the technocratic and administrative elite. They were kind enough to take time out of their work to talk to me, and they provided me with a network of personal introductions that rapidly led me from offices in Mexico City to SEMARNAP offices in Oaxaca.

From Mexico City, I moved onward to Oaxaca, attending regular meetings of regional forest councils, where forestry officials encountered members of indigenous forest communities. I tried to talk also to forestry officials and foresters, visiting their offices and asking them what kind of work they did and how they did it. Finally, I spent six months of intensive fieldwork in the community of Ixtlán de Juarez, accompanying logging technicians into the field, talking to loggers and community elders. Such multisited ethnographies impose their own challenges: Perhaps the continual doubt that I should have been elsewhere, that I was not in any one place long enough to understand what was going on, is analogous to the rootless cosmopolitan's fantasy about local belonging and the attractions of being in one place.

2

Building Forestry in Mexico: Ambitious Regulations and Popular Evasions

Over the last 200 years, governments all over the world have taken up the burden of knowing, managing, and protecting nature, accompanied by the task of developing economies and caring for citizens. Most citizens, of First or Third World countries alike, now take it for granted that the state is responsible for preventing environmental degradation and developing natural resources in the national interest. This is a relatively new event, a massive expansion of state presence that is manifested through towering glass and steel office buildings in capital cities or more modest offices in state capitals, through government technicians who travel through the countryside, and through the publication of national statistical reports on such matters as deforestation, timber production, or areas of forest fires. The state has become accountable as a certain kind of thing, as a more or less enduring, solid, and unitary knower of an environment that it is supposed to act on and protect. In this sense, the history of forestry in Mexico over the last century is similar to the efforts of other nation states around the world: governments from the United States (Worster 1979) to India (Sivaramakrishnan 2000), attempted, within their means, to control, develop, and protect agriculture and forests. New state institutions drew on internationally circulating scientific theories about forests and climate, scientific forestry (silviculture), and representations of fire and fire users as barbarous, dangerous, and uncontrollable. It would be easy then to see the Mexican case as yet another instance of a global environmental discourse arriving in a particular place, an overly familiar story, too often told. The local details of history and culture, the precise details of who was president and which official lost his job, may seem irrelevant to readers who are concerned with distant countries or not particularly interested in forests and the environment. Environmental policies move rapidly: Forestry is no longer the cutting edge of state environmental science, and readers may be more interested in biodiversity protection or perhaps with the promise that forests may store carbon and help reduce climate change. In this chapter, I will try to convince you that thinking about how forestry science came to Mexico and came to drive a regime of bureaucratic control is important to anyone who is interested in how states, bureaucracies, and technical knowledge come to affect each other. New regimes of bureaucratic control and technical knowledge are proposed for emerging technical and environmental problems of all kinds, from the problem of global climate change to the recent oil spills in the Gulf of Mexico. How do such regimes get made and how do they acquire their stability? How does a scientific regime become embedded in collective imaginations and stabilized as a public fact and with what kinds of consequences?

The particular history of Mexican forestry institutions helps us understand how making scientific knowledge and state-making go side by side, how the uneven, patchy, and particular spatial expansion of the Mexican forest service has brought relatively weak officials into encounters with indigenous forest dwellers who speak for relatively stable, enduring, and legally recognized forest communities. The history of Mexican forestry, therefore, has much to tell us for the larger questions of how states at other places and times have tried to tame knowledge and perform their authority, and to help us understand how a globally mobilized scientific knowledge is translated from the laboratory into distant places. Scientific knowledge does not travel smoothly into people's daily lives: As we shall see, the political culture of Mexico, with its history of revolution, land reform, and environmental crisis, has powerfully influenced how scientific knowledge about forests and public understandings of the Mexican state have come to be produced. The details of government documents and reports, the precise content of forestry regulations, and the histories of government offices and model forests have influenced the credibility of official knowledge and the authority of the state. Particular places and documents, offices, and biographies are not necessarily small: On the contrary, it is by asking how these particulars are folded into what we ordinarily take to be large-scale categories that we can begin to see why local environmental and institutional histories may affect the destinies of states and of scientific knowledge.

Mexican forestry starts from particular places: humble buildings, particular government offices, and geographical locations. These beginnings have powerfully affected what it means to imagine and claim control of forests in the present. One of the beginning places for forestry in Mexico is the campus of the University of Chapingo, the center from which Mexican forestry science was brought into the Mexican state and a place with which many forestry officials still identify. Formerly one of the key members of the Aztec empire's triple alliance, Texcoco is now a medium sized provincial town about an hour's drive from Mexico City. In 2000, I traveled to Chapingo, emerging from the sprawl of the city to drive across the desiccated former lakebed, passing drainage canals, dry garbage dumps, rusting machinery, and plantings of dust stabilizing reeds. The main entrance of the university was on a busy highway; a noisy stream of buses and trucks passed directly before an imposing arch that framed an avenue of trees that lead to a quieter interior. Modernist concrete buildings were scattered across a large campus; at the core was the original *hacienda* building, where the national school of agriculture was founded in 1926. During my visit, I was given an unofficial tour of an ordinarily closed hall: Poorly lit murals by Diego Rivera celebrated the Mexican state as restorer of indigenous agricultural knowledge and as the bearer of technological and scientific progress. By the time I arrived, Chapingo no longer felt like such an avatar of progress, and Rivera's optimism felt like a hollow promise. Like the state universities I knew in the United States, the campus had a sense of grand aspiration and of current neglect, of modernism left behind by history. The buildings were rather tired looking, even as the gardens were lovingly kept and well watered. The entrance to the forestry department contained glass exhibit cases of forest pests and tree samples, notices of examinations to be taken, and a cross-section of an enormous ancient tree, the kind of thing that seems to be obligatory in forestry schools around the world. When I visited the libraries, I found the same sense of former prosperity and present decline. Journals that had been initiated in the optimistic glory days of the 1970s oil boom had ceased to exist, and the catalogs and shelves that I searched had few recent books.

The feeling of decline at Chapingo was partly due to the fortunes of the Mexican economy, but there were other reasons for this sense of abandonment. Since the early 1980s, repeated economic crises have undermined the state's ability to support the university system. However, as I gradually learned, Chapingo had also been sidelined by shifting environmental politics and by the dramatic retreat of the state from industrial forestry. The forestry professors I met at Chapingo complained bitterly of the declining prestige of agriculture and especially forestry, which they blamed on the rise of alternative projects of biodiversity mapping, conservation, and nature protection, sponsored by alternative universities and state institutions. In the 1970s, Chapingo had been one of the places where modern Mexico was to be produced, where the technicians of state-sponsored forest industrialization were to be trained. With repeated economic and political crises and with the declining prestige of the Mexican state, the government has withdrawn from direct control of forest industrialization. Mexican environmentalists have pushed for alternative projects of biodiversity protection, ecology, and conservation, leading to the formation of new state institutions such as the National Institute of Ecology, with its academic allies among biologists in the National Autonomous University of Mexico City (UNAM). Idealistic young Mexicans who want to work with nature are now more likely to seek training as biologists who protect forests than as foresters who log them.

At the high point of industrial forestry in the mid-twentieth century, cutting trees for timber production had become part of a state project of modernizing and developing Mexico. By the time I arrived in 2000, many people felt that this was no longer a promising social or environmental future. Three particular aspects of scientific forestry were translated to Mexico in the early twentieth century and subsequently incorporated into projects of controlling forests and society. Forestry initially came to Mexico through its promise to make connections among forests, climates, and floods in the scientific theory known as desiccation theory, persuading urban elites to support the expansion of the state into forests. The practices of scientific silviculture came to define the professional identities of foresters and officials who wished to tend forests, produce timber, and make forests legible to the state. Finally, images of forests as threatened by fire and by fire using rural people came to become vitally important in stabilizing forestry institutions that were opposed by unruly indigenous or peasant agriculturalists. These theoretical and practical projects were first advocated by scientific entrepreneurs in Mexico City in order to gain the attention of policymakers and urban elites who were persuaded to support the creation of a national federal forest service. Subsequently, forestry regulations and scientific theories had to be translated into documents that traveled to recalcitrant provinces and distant forests, often with uncertain results.

A Brief History of Mexican Forests

When I started working in Mexico, I had to explain to friends that, yes, Mexico does have forests (about a third of land area, more or less the same as the United States). For many outsiders as for many Mexicans, narratives of deforestation and degradation are so powerful that this comes as something of a surprise. In fact, many towns in Mexico are within sight of the forested mountains where most of the country's forests are located, the marching ranges of pine and pine-oak forests that resemble parts of the American West, or perhaps of Mediterranean Europe. Even from Mexico City, the rare clear day will blow aside the haze of atmospheric pollution to reveal the astonishing sight that this enormous city is surrounded by heavily forested mountains. However, the impression of a bare landscape is not entirely incorrect: Much of northern Mexico is indeed desert or semi-desert covered by grassland and scrubby landscapes; it is the mountain ranges on the east and west of the country (see figure 2.1) that are covered by pine, pine-oak, and fir forests between about 2,000 and 3,500 meters above sea level. The tropical lowlands on each coast were formerly covered by dense tropical moist forests, which have now largely been cleared for agriculture and ranching. Popular and academic narratives of ever-increasing

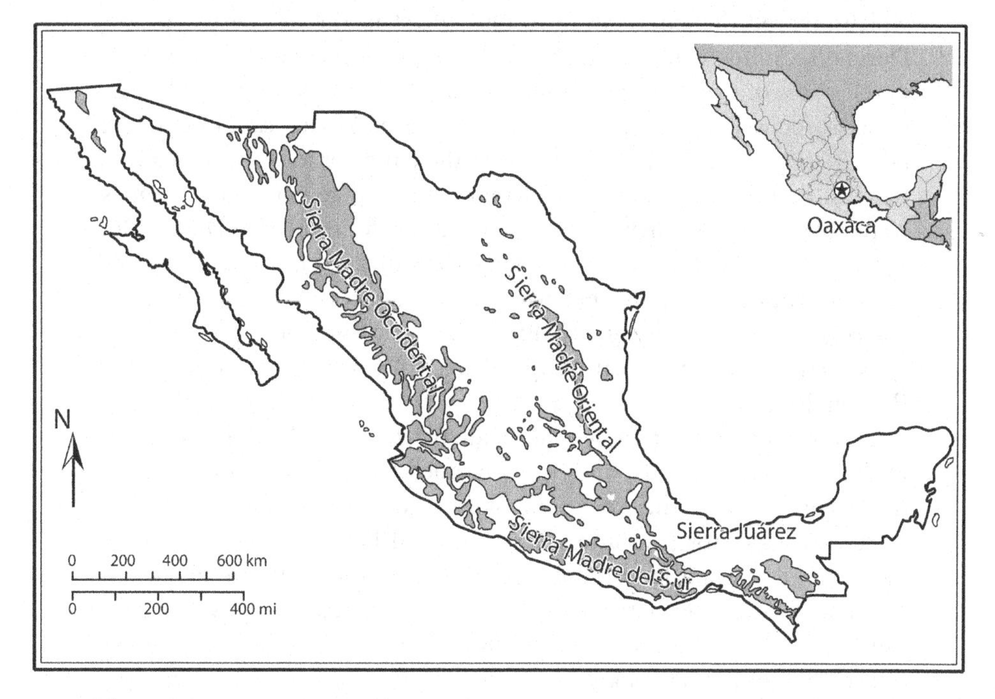

Figure 2.1
Original distribution of pine-oak forest type in Mexico. *Source:* Map by Metaglyfix, drawn from Comisión Nacional de Biodiversidad after (INEGI 1990b).[16]

deforestation and environmental degradation obscure a much more dynamic history; in fact, forests increased dramatically when agricultural land was abandoned during the catastrophic population crash that followed the Spanish conquest in the sixteenth century. After an initial flurry of interest in controlling grazing and logging, the Spanish crown remained largely uninterested in directly controlling or protecting forests, beyond assuring local municipalities of their timber and firewood needs through legal title to surrounding forests. Large-scale logging of more distant forests did not begin until the mid-nineteenth century, when expanding railroads, capitalist investment, and agricultural commodity exports caused the rapid clearance of large areas of tropical forest in the lowlands and the logging of pine timber in the highlands.[1]

During the nineteenth century, the Mexican state had been unable and unwilling to assert direct control over Mexican forests, preferring to allocate control and management of forests to large private land buyers and concessionaires. By the outbreak of the Mexican Revolution in 1910, vast tracts of forest land had been sold or titled to foreign and national investors who had benefited from the policies of President Porfírio Díaz, which favored the exploitation of pine forest in the northern states of Durango, Chihuahua, and the extraction of precious tropical hardwoods in the states of Chiapas and Campeche (de Vos 1996; Lartigue 1983). In the more densely inhabited central states such as Michoacán, forests had been fragmented into smaller parcels in the hands of logging companies, but this had occurred in recent memory, and indigenous municipalities were able to reclaim their forests after the revolution (Espin Diaz 1986; Vazquez Leon 1986). Overall, in many areas of the temperate highlands and the tropical lowlands, extensive areas of forest were economically inaccessible and were largely untouched, except for firewood and construction uses by rural people (e.g., in Oaxaca and Campeche) (Haenn 2002).

Forests had been very much the residual category in land use struggles in the nineteenth century, and this continued to be the case in the twentieth century. With the explosion of the Mexican Revolution (1912–1920), struggles for restitution and control of land were often over the meaning of colonial land grants and the usurpations of the nineteenth century.[2] In the aftermath of the revolution, one-party rule was established by the Institutional Revolutionary Party (PRI), which ruled uninterruptedly until 2000. A key plank of the PRI/state's legitimacy was the state's claim to have delivered land to the landless, whether by redistributing the enormous holdings of the great estates directly to newly formed peasant communities or by restoring land to municipalities that claimed ownership based on colonial era titles.[3] Antony Challenger (Challenger 1998) traces two main pulses of land titling: under President Lázaro Cárdenas between 1934 and 1940, when primarily agricultural land was handed to newly created ejidos and communities[4]; and between 1958 and 1976, when the state responded to rural unrest by titling most of the remainder of Mexico's forests. This state response tapped into the popular and state-sponsored revolutionary myth of agrarian reform (Joseph 1994a; Mallon 1994), while in fact delivering forest land that could not be used for agriculture. The result is that at present the vast majority of Mexico's forests are communally owned.[5] The political status of indigenous people and communal land tenure remain an intractable reality with which the Mexican state continues to struggle, resulting in swings between increased community autonomy and increased state or private intervention into forests. As we shall see in the following chapters, the political and environmental histories of indigenous forest communities have placed continuous pressure on the authority of the forest service and the Mexican state (see figure 2.2 for a summary of political and economic events).

Over the last century as at present, industrial logging has taken place mainly in the pine oak forests of the temperate highlands along the Sierra Madre Oriental and Occidental: Currently, more than 95% by volume of industrial timber is pine timber cut in these areas (FAO 2001). This was partially due to the historic importance of the highlands, which is where the principal cities and population centers are located. In addition, however, pine oak forests contain a high density of commercial timber compared with the tropical forests of the Atlantic lowlands. Although logging of precious tropical hardwood timbers such as mahogany has historically been of great importance in the tropical forest areas of the lowlands (de Vos 1996), valuable species such as mahogany are found at densities of perhaps one tree per hectare (Snook 1995), making total timber production from these areas small in volume. By comparison, pine timber is found in forests where one or two pine species dominate any given area, making the project of scientific forestry, and of logging, transport and processing cheaper and easier.

Revolution and Environmental Order

The Mexican Revolution of 1920 dramatically upset the political and economic order of the country. In many parts of the country, rural people

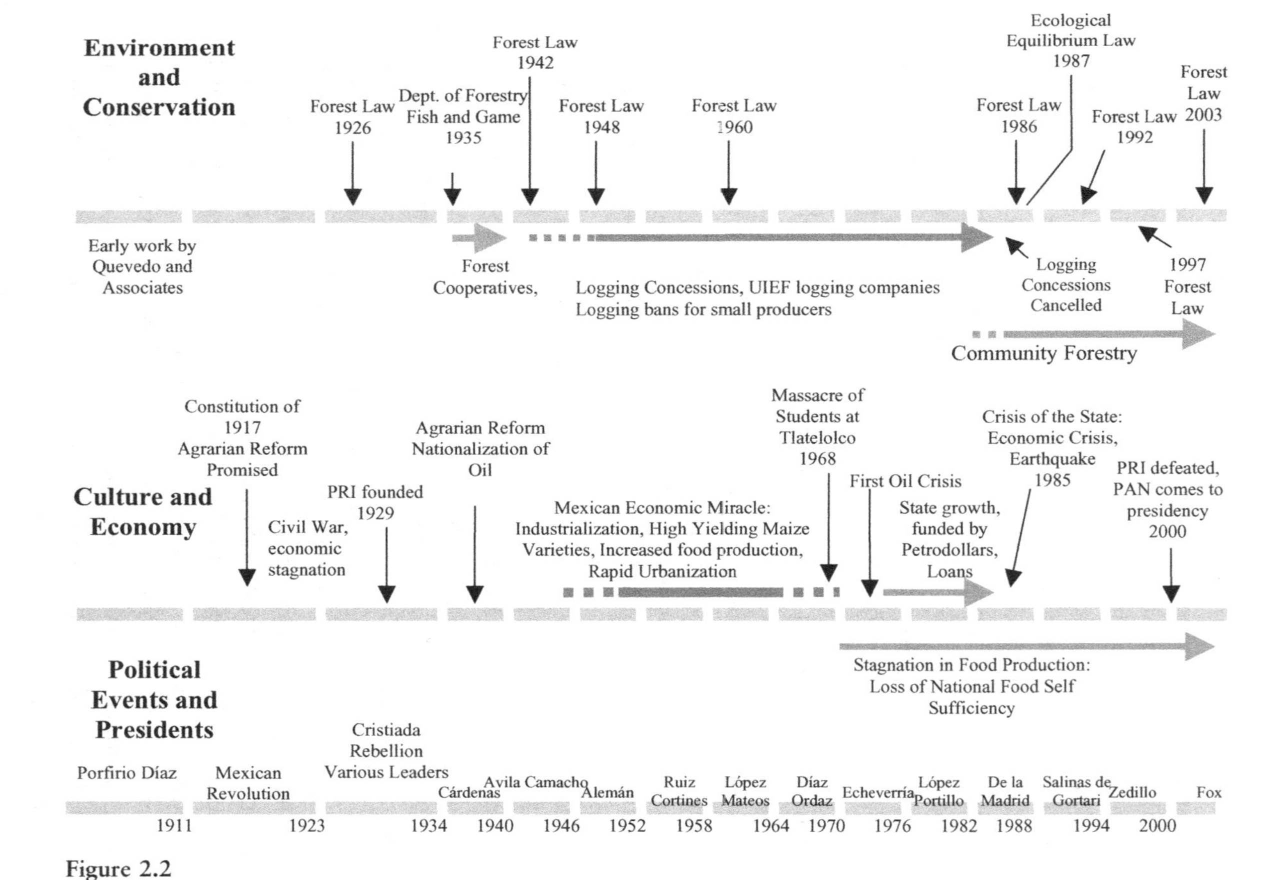

Figure 2.2
Timelines of political, cultural, and economic events.

fled to the hills to avoid the marauding armies; crops were abandoned; bridges, factories, and buildings were destroyed; and millions of people died from warfare, disease, or starvation. By the end of the main fighting in 1920, the economy was in ruins. The postrevolutionary Mexican state was therefore vitally concerned to reestablish order, including through a dramatic expansion of the territorial authority of the federal government in the forest law of 1926. A modest office for forests was later expanded into the department of forests, hunting, and fisheries (*Departamento Forestal y de Caza y Pesca*) in 1934.[6] The goal of the new forest service was to create a rational/legal forest administration that would enforce a detailed web of forest regulations over the whole of Mexico's forests. Officials tried to protect forests in the name of watershed protection, in order to prevent the ravages of forest fires they believed to be set by poor rural people, and to organize and tend forests through scientific silviculture. Between 1926 and 1940, the fledgling institution acquired the offices, employees, forest guards, and forms required by a proper bureaucracy. The new institution claimed to have authority and responsibility "so as to attend with greater effectiveness to the affairs connected with forest administration in *the whole extension of the national territory*" (Quevedo 1938:39, my emphasis).

Mexican Environmentalism and the Valley of Mexico

Institutions do not spread uniformly over landscapes: They are affected by the particular geographical and historical circumstances of the place and time where they come into being, as by the kinds of resistances that they encounter. In the case of the forest service, the location of Mexico City made watershed protection of particular concern to officials and politicians who lived in a dry intermontane basin surrounded by forested mountains, and who were exposed to floods, water shortages, and polluted air. Over the last century, the symbolic and material power of the state have resulted in the massive growth of Mexico City, the drainage of the lake of Mexico,[7] and increasing air pollution that is trapped by surrounding mountains. State centralization has ensured that the majority of forest service officials live and work in the valley of Mexico, making local environmental history and landscapes a resource for environmental advocates who have tried to influence official environmental theories and policies. Flood protection and climate change were of great local importance as a history of floods had driven the colossal state project of draining the lake of Mexico in the late nineteenth century

(Connolly 1997; Gibson 1964; Musset 1991; Quevedo 1926). This local concern with floods and climate was to inspire forest protection in the whole of Mexico for the first half of the twentieth century.

By the late nineteenth century, a group of engineers, meteorologists, and scientists based in Mexico City were concerned about the impact of deforestation on climate, water supplies, and flooding, making use of the then received scientific theory of *desiccationism* (see box 2.1). The leading light of this circle was the engineer and public intellectual Miguel Angel de Quevedo.[8] Over forty years of publication and advocacy, Quevedo produced a coherent scientific discourse of public reason and environmental degradation, deploying visual representations of degraded landscapes and poor farmers to represent the valley of Mexico as degraded

Box 2.1
Desiccation Theory: Does Deforestation Cause Droughts, Floods, and Declining Stream Flow?

Many people intuitively feel that cutting down forests will have some effect on climate and water supplies. Far from being a fact that is well established by contemporary science, this is a hotly contested theory with ancient roots. According to desiccation theory, deforestation causes declining rainfall, increased flooding after rainstorms, and long-term declines in stream flow. This theory of desiccation makes distant forests of great interest to states and citizens concerned with water supplies or climate. Scientists as far back as Aristotle and Theophrastus in ancient Greece make mention of desiccation theory, but it was only during the eighteenth century that it was formalized as a coherent scientific theory in the tropical island colonies of the French, Dutch, and British empires, becoming the main environmental justification for forest protection by colonial powers during the nineteenth century (Grove 1995), and by nation states in the first half of the twentieth century (Bryant 1996; Saberwal 1997). In Mexico scientific societies had been concerned about the possible effect of deforestation on the climate as early as the mid-nineteenth century (Simonian 1995), and desiccation theory was so generally accepted during the late nineteenth and early twentieth centuries (Conzatti 1914; Serrato 1931; Vasconcelos 1929) that it has come to be internalized in current popular understandings of deforestation, often forming the basis for popular opposition to logging or road building (Mathews 2009). In general, environmental scientists consider classic desiccation theory to be only partially correct: Around the world, there are heated controversies about its scientific validity (Bruijnzeel 2004; Calder and Aylward 2006; FAO and CIFOR 2005) and over the political, cultural, and economic effects of policies justified by desiccation theory (Forsyth and Walker 2008).

and in need of restoration, protection, and rational scientific management (de la Vega 1933; Quevedo 1926, 1933). Quevedo argued that environmental restoration should be carried out by the state and directed by scientifically trained elites, and he attributed degradation largely to indigent rural people who burned fields and set fire to forests, negatively affecting climate and water supplies. Further, the expanding city needed parks and green spaces in order to ensure the health and psychological well-being of urban residents (Quevedo 1935[1910]); Wakild 2007).

Quevedo and his associates believed that poor peasants were burning forests in order to cultivate steep slopes, where bare soil would soon be eroded away. In earlier writings, Quevedo depicted rural people as being poor and ignorant peasants: Indigeneity was a relatively unimportant political category at this time, and it was only in occasional later writings that Quevedo criticized specifically *indigenous* agricultural burning (Quevedo 1941) carried out by ignorant indigenous people who believed that rising clouds of smoke would summon the rains (Quevedo 1928). In later years, rural indigenous people became the imagined opposite of modern state knowledge. Although for Quevedo indigenous people were less important, he did firmly cement in place an official discourse where the beliefs of rural people, whether peasants or indigenous, were the polar opposites of secular scientific knowledge and reason. Taken as a whole, the publications of Quevedo and his circle were skillful rhetorical and visual performances of public reason and rural ignorance. They succeeded in linking policies of state environmental control in distant forests with the concerns of urban audiences who were quite willing to believe in the barbarism or ignorance of distant rural people. Representations of rural ignorance and official scientific knowledge were to drive state projects of environmental control for the rest of the twentieth century.

Quevedo had already had a notable influence on the 1926 forest law, but it was during the presidency of Lázaro Cárdenas (1935–1940) that he succeeded in securing national attention and state support for an expanded forest service (see figure 2.3). In a radio speech to the nation in 1935, Cárdenas announced that forest resources were necessary for national economic development and that forest protection would restore climactic equilibrium and the flow of streams and springs (Cárdenas 1935). Quevedo was appointed head of the new Department of Forest, Hunting, and Fishing in 1935, where his principal efforts were concentrated on preventing forest fires, preventing logging in sensitive watershed forests, and writing and applying detailed logging and timber transport regulations (Quevedo 1938; Calva Téllez et al. 1989). Environ-

Figure 2.3
Political and technical authority meet: President Cárdenas (left) and Quevedo (right), 1937. *Source:* Courtesy of the Luz Emilia Aguilar Zinser Archive.

mental education campaigns promoted fire fighting and understandings of forests as protectors of water supplies. Logging was restricted or entirely banned in forest protection zones (e.g., the forests of the valley of Mexico [1933] and the watersheds of irrigation projects in the whole country [1934]), while model forest management plans were published in journals (e.g., Atlamaxac in Puebla) (Treviño Saldaña 1937), visually demonstrating the kind of technical order that was desirable. These initiatives aimed to produce and control specific new places: model forests, streamside protection areas, and degraded lands. These places were geographically defined objects; rather than being uniformly spread in space, state claims over forests were patchy, supported by documents, local management plans, and model forests. These new objects were in turn allies and resources that supported official efforts to represent the state as a source of order and science, and rural people as sources of disorder, degradation, and ignorance (see figure 2.4). The state was being performed as an entity that could know and manage forests across the entire country; this performance was supported by a careful weaving together of eminently local natural objects, official reports, documents, and regulations. It was this web of alliances that supported official knowledge

Figure 2.4
Ideal disorder: image of a farmer outside Mexico City, 1933. . . . the [corn] plants barely reach 60 cm. in height, with no ears. In the background . . . are vestiges of the forests . . . which are still to be cut, in a conversion to cultivation which is contrary to common sense. (de la Vega 1933). *Source:* Courtesy of México Forestal, *Journal of the Sociedad Forestal Mexicana*, founded by Miguel Angel de Quevedo.

and the gradual expansion of the forest service from Mexico City to state capitals and distant forests.

Quevedo drew on the latest scientific theories to justify the new forest service: These included the international scientific theories of desiccation (box 2.1), understandings of forest fires and agropastoral burning as destructive (box 2.2), and the silvicultural science of tending forests and producing timber (box 2.3). Quevedo was familiar with the latest international scientific thinking of French forestry scientists and with the organization of the French and American forest services (Simonian 1995:67–84). His goal was to combine administrative rationality and scientific knowledge about nature so as to conserve *and* exploit forests. Logging companies and forestry cooperatives alike were to be supervised and guided by the state. Through the writings of Quevedo and his circle, a particular conception of nature and culture were institutionalized, according to which forests and agriculture were conceptually and legally

Box 2.2
Burning Forests: Technological Intervention or Barbaric Destruction?

"Swidden" is the term for a vast variety of long fallow agricultural systems found around the world. In these systems, forest vegetation of varying ages is cut and burned for fertilizer (hence, the derogatory term "slash and burn agriculture"), and agricultural crops are grown for a few years, after which the field is abandoned to regrow to forest. After a number of years of fallow, the agricultural cycle is reinitiated. Governments around the world have typically demonized swidden as an unspeakable and destructive "other." There is abundant evidence that the various forms of swidden agriculture practiced in Mexico (milpa, tlacolol, coamil) are not necessarily destructive. The literature on swidden agriculture in Mexico and elsewhere shows that it has been successfully practiced for many centuries and that it is a sustainable agroecological technology when population densities are not so high as to cause overly short fallow periods (Dove 1983; Alcorn 1984; Whitmore and Turner 2001; Nigh 1975). In Mexico the ministry of agriculture has attempted to support permanent and sedentary agriculture, making swidden agriculture dangerous as it migrates across conceptual and institutional boundaries. In this sense, swidden is a "monster" (Bowker and Star 1999) that is marginal and unknowable and that threatens agriculture (because it is unmodern) and forests (through agricultural clearance). State institutions are all important in entrenching these categories and marginalities: If a "Ministry of Swidden" existed anywhere in the world, we would no doubt read concerned reports about the threat of "overmature forests" and "tired" agricultural land.

separate, and agricultural fires that threatened to transform forests into fields were destructive and abhorrent. This aversion to fire and agriculture is typical of forest services around the world, but in Mexico this had particular ecological and political consequences. The state regarded forest fires as completely unnatural even in pine and pine-oak forests where fire was essential for forest regeneration. Further, fires were being used by indigenous people whose cultural difference was and remains conceptually indigestible and politically intractable.

Quevedo's attempts to enforce forestry regulations and limit agricultural expansion soon encountered opposition from the ministries of agriculture and agrarian reform. Most importantly, protecting forests conflicted with President Cárdenas's project of cementing the power of the Mexican state by giving land to peasants. In 1940, Cárdenas closed the Department of Forestry Fish and Game and transferred responsibility for forests to the Ministry of Agriculture, which was concerned with

Box 2.3
Silviculture: The Science of Tending Forests

> Silviculture is the set of theories and practices through which foresters connect forest ecology with logging and other management practices, with the goal of directing the long-term development of forests. Silviculture has been one of the principal theories through which foresters identify themselves as professionals who know, manage, and protect forests. "Silviculture has been variously defined as the art of producing and tending a forest; the application of knowledge of silvics in the treatment of a forest; the theory and practice of controlling forest establishment, composition, structure, and growth" (Smith 1997). In the first half of the twentieth century, late colonial and early postcolonial states around the world codified silvicultural systems that prescribed logging practices (Buschbacher 1990), often through control of the minimum diameter of tree that could be cut. Critics of Mexican forestry have often claimed that logging companies have destroyed forests through clear-cut logging. Genuine clear cuts that remove all standing trees and scarify mineral soil have in fact almost never been applied, as this is more expensive than "high-grading" when only large and valuable trees are cut, leaving damaged or undesirable trees but preserving forest cover. A 1932 forestry ordinance allowed relatively heavy seed tree and clear-cut systems to be applied (Calva Téllez et al. 1989:169), but a 1943 regulation restricted logging to 35% of standing volume in previously unlogged forests, with a minimum cutting diameter of 40 centimeters (Musalem 1972). This regulation was in theory uniformly applied to all of the temperate forests of Mexico until 1980 (Musalem 1972). It was not until 1982 that the Metodo de Desarrollo Silvicola (MDS) was developed (Rosales Salazar et al. 1982), allowing heavier cutting in order to promote pine regeneration.

expanding agricultural lands rather than with protecting forests. There was to be no independent forest department until 1951 (Mejía Fernández 1988). Subsequent to Quevedo's loss of power, desiccation theory was attacked by agronomists whose central focus was agriculture and who supported land reform and agricultural development (Contreras Arias 1950a, 1950b). From the 1940s onward, the principal role of forests from the point of view of the state became erosion prevention and flood control; desiccation passed from being a scientific a theory that justified state control of distant forests to a popular environmental theory used by rural communities to protest against industrial logging and to find allies among urban audiences (Mathews 2009).

Quevedo's dismissal and the subsequent reorganization of the forest service illustrate the patterns of personal career instability and institutional fragility that have prevailed over the last eighty years. The forest service, in its various incarnations, has been plagued by career instability for officials, frequent reorganizations and renamings, and a constant succession of new forest laws (see figure 2.2 and table 2.1). This instability is in striking contrast to the enduring stability of discourses of fire prevention and the continued commitment to rational and technical forest management. Officials continue to fight forest fires and prevent agropastoral burning, forests and fields are conceptually and bureaucratically separate, and silvicultural science is used to manage and order forests that are regulated through elaborate systems of documentation and regulation. The bureaucracies concerned with agriculture and agricultural reform have consistently been able to overrule the forest service in practical terms, but apart from desiccation theory, the discursive categories and technical knowledge established by Quevedo have remained unchanged.

Regulating Forests

The enormous administrative task that Quevedo faced when he took office can be grasped if we compare the number of available trained

Table 2.1
Institutional Location and Name(s) of the Federal Agencies Responsible for Mexican Forests, 1911–2008

Years	Name(s) of federal agencies responsible for forests	Institutional location of forest service and comments
1904–1908	Junta Central de Bosques y Arboledas (Forests Committee)	Ministry of Public Works
1908–1912	Departamento de Bosques	Agricultura y Fomento (Agriculture and Development)
1912	Depto. de Conservación de Bosques (Forest Conservation)	Agricultura y Fomento (Agriculture and Development)
1920–1929	Dirección Forestal y de Caza y Pesca (Forests and Fisheries)	Agricultura y Fomento (Agriculture and Development)
1927–1929	Direc. Gral. Forestal y de Caza y Pesca (Forests and Fisheries)	Agricultura y Fomento (Agriculture and Development)
1929–1934	Departamento Forestal	Dirección de Fomento Agricola (Agriculture Ministry)
1934	Depto. Forestal y de Caza y Pesca (Forests, Hunting and Fishing)	Independent Forest Department

Table 2.1
(continued)

Years	Name(s) of federal agencies responsible for forests	Institutional location of forest service and comments
1940–1951	Dirección Gral. Forestal y de Caza (Forests and Hunting)	Ministry of Agriculture-(forests are low-level agency as Dirección General)
1951–1960	Subsecretaria de Recursos Forestales y de Caza (Forests and Hunting)	Ministry of Agriculture, (forests promoted to Subsecretariat)
1960–1982	Subsecretaria Ftal. y de la Fauna (Forests and Wildlife)	Ministry of Agriculture
1982–1985	Subsecretaria Forestal (Subsecretariat of Forests)	Ministry of Agriculture, Wildlife moved to another agency
1985	Subsecretaria de Desarrollo y Fomento Agropecuario y Forestal, Dirección Gral. de Normatividad Forestal	Forest service demoted and divided between two different agencies in Ministry of Agriculture. (Economic crisis year)
1986	Comisión Nacional Forestal (Comisión consultora integrada por la SARH, SEDESOL y la SRA)	National Forest Commission made up of Ministries of Agricultural Reform, Social Development, and
1986	Subsecretaria de Desarrollo y Fomento Agropecuario y Forestal (Subsecretariat of agricultural and forest development)	Irrigation and Agriculture (SARH)
1988–1995	Subsecretaría Forestal (Subsecretariat of Forests)	SARH
1995	Subsecretaría de Recursos Naturales—SEMARNAP	SEMARNAP
1997	Secretaría de Recursos Naturales (Subsecretariat of Natural Resources)	SEMARNAP (Ministry of Environment Natural Resources and Fisheries)
2000	Subsecretaría de Gestión para la Protección Ambiental, Comision Nacional Forestal (Environmental Protection, National Forestry Commission)	SEMARNAT (Ministry of Environment Natural Resources)

foresters with the vast extent of Mexico's forests. By my calculation, he could at best have deployed about one forester for every half million hectares of forest. A critical limiting factor for the new forest service was therefore the sheer lack of professionals who could implement regulations and fill in documents. Quevedo had helped set up a forestry school in Mexico City in 1911 (the Escuela Nacional Forestal), which moved to Coyoacán in 1916 (see figure 2.5 for an image of the class of 1918), and then to Chapingo near Texcoco in 1926 (Arteaga 2000, 2001). He had imported French forestry professors to teach forestry and silviculture (Junta Central de Bosques 1911, 1912), so that by 1927, 282 forestry professionals had graduated (Borgo 1998:252). Some of these did not choose to work as foresters, and others had been killed during the revolution, making the actually available number still smaller. By 1939, the total size of the forest service was approximately 1,500 people,[9] of whom at most 200 or so were foresters out of perhaps 500 administrators, with more than 1,000 forest guards (Hinojosa Ortíz 1958). This was still a small number with which to control the forests of an entire country.

Figure 2.5
Students of the Escuela Nacional Forestal in Coyoacán, 1918 (Peña Manterola 1967). Note the military-style uniforms worn by the cadets. As late as the 1950s, forestry students received a distinctly paramilitary training. *Source:* Courtesy of México Forestal, *Journal of the Sociedad Forestal Mexicana*, founded by Miguel Angel de Quevedo.

Dangerous Papers: Transport Documents and Burning Permits

The tension between the small numbers of personnel and the dramatic reach of regulations can be illustrated by considering logging and fire regulations. Burning in theory had been severely controlled since 1926, with additional regulations in 1932 (Gutiérrez 1930; Dirección Forestal y de Caza y Pesca 1930, 1932; Mares 1932). In practice, the forest service never had the manpower to enforce these regulations; fire is used for agriculture and pastoralism in large parts of Mexico, and it was impossible for officials to prevent rural people from burning. If fire control regulations had been enforced, tens of thousands of farmers and pastoralists would have had to apply for permission to burn in writing (Calva Téllez et al. 1989), deluging the offices of overworked officials. Rather than controlling actual burning, therefore, fire regulations created a vast field of criminal behavior, which officials might punish if they had the political strength and material means to do so. People who chose to burn learned to do so either discretely or in secret, and officials learned to turn a blind eye.

Like fire control regulations, logging regulations asserted an ideal order, an abstract and coherent universe within which trees were documented almost from their germination to their sale as saw timber or furniture. Timber was only supposed to be cut if it was on land with a forest management plan. Plans had to have increasing detail and rigor as the area of the exploitation increased, from simple tickets for subsistence firewood gatherers to full-scale management plans for exploitations of more than 5,000 cubic meters of timber (decree of 8/9/1930). Even the minimal requirements of this decree were probably impossible to carry out; firewood gatherers and charcoal burners had great difficulty obtaining the required papers, and, as we shall see, they largely evaded official controls. All timber had to be marked by a forester before it could be cut; the forester was supposed to send the forestry department a monthly report on the volumes of timber he had marked. Timber was supposed to be transported with documentation proving that it had been cut legally (*Circular* 83–35 of June 1935), but this regulation must have had little success because it was soon followed by another regulation (*Circular* 97–35 of August 1935) ordering that only legally obtained forest products could be documented, and then by *Circular* 5–38, which forbade the issuing of logging permits without transport documentation (Calva Téllez et al. 1989:393–441). Regulation followed on regulation, as Quevedo struggled to control the practices of forestry officials and logging companies, who were adept at manipulating the documents that his regulations demanded. Transport documentation had become a valuable commodity because it could protect timber from official sanctions (and probably from requests for bribes). As a former head of the forest service pointed out a few years later, "Forestry documents have acquired an autonomous value, they are traded and speculated" (Hinojosa Ortíz 1958:40). Far from producing official knowledge, then, documents produced obscurity and the possibility of trading and producing value.

Foresters and officials who worked in the years immediately after the Quevedo administration recalled that their work consisted mainly of making sure that pieces of paper moved through the system smoothly. The most important papers were the voluminous timber transport documents, not only because they allowed timber to be bought and sold but because officials used these documents to produce national timber production statistics that performed the state as knowledgeable and in control of forests. The forest service was highly centralized; management plans were approved in Mexico City, and forestry officials were supposed to approve plans for forests that they had never seen on the basis of the consistency and completeness of the documents before them. A former head of the forest service pointedly observed that:

> It can be affirmed that the Forest Service has a city based personnel, living permanently in the national and state capitals, and that it lacks rural personnel that stays in constant contact with the forested areas. (Hinojosa Ortíz 1958:32)

This pattern too has been consistent over the last decades: Designers of new forestry regulations have rarely considered the material inability of city-based forest service officials to visit forests and oversee management practices; the very possibility that a centralized bureaucracy could control forests has almost never been called into question. Officials' constant reaffirmation of the importance of documentation has literally created the Mexican state in the minds of officials and rural people alike. Just as rural people learned about forest policies through their efforts to secure documentation or avoid regulations, forestry officials understood forests through the documents that they spent most of their time dealing with.

Forest Industrialization as Progress, 1940–1972

In keeping with his commitment to land reform and rural communities, President Cárdenas set up forestry cooperatives across Mexico between

1935 and 1940. Community forests were supposed to be logged only through forest cooperatives, but as with the agricultural communities created at this time, the state lacked the financial resources to support the forestry cooperatives. Instead, many private logging companies would sign contracts with paper "cooperatives," often formed with the legal and financial assistance of the companies themselves (Hinojosa Ortíz 1958:91).

After Cárdenas, community forestry fell into disrepute, and the forest cooperatives were closed down by the late 1940s (Calva Téllez et al. 1989). With the beginning of the Second World War, logging boomed: Railroads, mining, and construction industries were hungry for timber. By this time, logging bans (*vedas*) covered large areas of the country, and most forests were officially closed to small timber companies between the 1940s and the mid-1970s (Calva Téllez et al. 1989:393–409). In addition to state-level logging bans, there were numerous local bans in national parks, near cities, or in watershed protection areas. In place of cooperatives the government chose to promote the formation of "industrial logging units," the *Unidades Industriales de Explotación Forestal* (UIEF) in the most productive forests in Mexico, effectively marginalizing the remaining small logging companies.[10] These large companies would be awarded logging concessions over an area of forest for an extended period of time, ranging from 25 to 60 years (Mejía Fernández 1988:38). The industrial loggers were supposed to pay landowners by the volume of timber cut as well as covering the salaries of government foresters and forest guards. Although these were notionally forest service employees, they inevitably came to answer to their direct employer rather than to the state.[11] With shifting political winds, most logging concessions were nationalized after 1960, but there was little practical difference and logging concessions, and working practices continued unchanged.[12] Throughout the concession period, the forest service continued to be highly centralized: Management plans had to be approved in Mexico City until the mid-1980s. Since then this responsibility has traveled to state-level offices, but there has been little transfer of budgetary authority so that the forest service remains in practice highly centralized.

For the state, the industrial logging concessions were a strategy for creating timber processing industries and establishing a state presence in pine forests, while providing a secure time horizon and a framework that would make forests legible to capital investments. There was probably only sporadic state presence outside of these commercial pine forests, but older foresters speak of this period as the golden age of Mexican forestry, describing a stable work environment and minimal state interference. For the first time, forest management plans were drawn up on the basis of aerial photographs (Escarpita Herrera 1959), and forest areas and timber volumes were beginning to be measured precisely. Officials approved of industrial logging concessions' ability to ensure that documents arrived smoothly and in good order; even if other regulations were flouted, at least timber volumes and areas were likely to be accurate. Alongside the industrialization of pine timber by large companies came an official neglect of firewood and charcoal production, which was carried out by *campesinos* and indigenous people and focused almost entirely on oak species. Thus, the mixed species pine and pine oak forests of the highlands were used by companies and rural people who almost entirely ignored each other: Industrial loggers focused on pines and rural people focused on oaks, even as pines and oaks were inextricably mixed.

The Decline of Industrial Forestry and the Rise of Community Forestry

Although they were the legal owners of most forests, communities received few monetary benefits from their ownership; the most important local uses were firewood cutting and charcoal burning, which were semi-legal, largely outside the market sector, and usually ignored by the state. During the 1970s, forest communities began to protest against the paternalistic and exploitative practices of the parastatal timber corporations in the highly productive pine forests of Michoacán, Oaxaca, and Jalisco (Espin Díaz 1986; Bray 1991; Chambille 1983). Forest communities in these areas wanted a larger share of the profits and jobs, and they began to demand direct control of logging and timber processing. The logging concessions had in part produced their opposition: They had created an infrastructure of roads and equipment that made the forest a valuable resource, while the labor organization of timber extraction created a group of local people who knew how to log forests and had learned scientific theories about forest management. From the late 1970s onward, forest communities began to form alliances with sympathizers and reformers within the forest service. The authoritarian/populist Echeverría presidency (1970–1976) responded to an initial wave of opposition to the logging companies by establishing 257 *ejido* forestry businesses and 25 unions of *ejidos* in north and central Mexico (Enriquez Quintana 1976; Mendoza Medina 1976). This was part of a larger policy of counterinsurgency: Land reform policies transferred vast areas of forest to community ownership during these years. New *ejidos* were increasingly aware of the value of their forests and ever more likely to demand greater control.[13]

Between 1983 and 1985, an unlikely coalition of forest communities, environmental nongovernmental organizations (NGOs), and reformist bureaucrats succeeded in pressing the Mexican state to cancel the logging concessions. The state faced a grave crisis of legitimacy as a result of the economic crisis and the Mexico City earthquake of 1985, neoliberal economic reforms were calling into question the legitimacy of state-owned companies, and the government was desperate to get rid of unprofitable and politically questionable logging companies. Newly formed community forestry businesses were largely protected by import taxes on timber until 1994 (Chapela 1997), ensuring relatively high prices that allowed some forest communities to gain dramatic increases in income from their forests (Snook 1995). In a select group of about thirty to forty well-endowed and politically powerful communities, the profits from logging have been reinvested in community forestry businesses that are integrated with traditional communal government structures (Bray et al. 2003). Through the successive forest laws since 1986, forest communities have been allowed to employ their own foresters and have become progressively less dependent on government technical advice and control. Nevertheless, forests must still be under management plans written by government-approved foresters and approved by government officials, who can still substantially determine when and how a community can log its forest.

Overall, the rapid succession of forestry laws since 1986 has strengthened legal community authority over forests, but there is still a wide range in the degree of practical control over forest management. Most communities sell their timber standing to outside contractors, giving community members little employment and the community little income. A small number of community forestry businesses employ their own foresters and operate sawmills, such as the community of San Juan Nuevo Parangaricutiro in Michoacán or the community of Ixtlán in the Sierra Juárez of Oaxaca, which is the subject of this book. National-level support for community forestry has waxed and waned since the mid-1980s. For example, between 1994 and 2000, the ministry of environment and natural resources (SEMARNAP) was generally supportive, and communities succeeded in gaining a significant share of forest subsidies.[14] The forestry laws of 1994 and 1997 created a legal structure that allowed long-term land leases, with the hope of encouraging international investment in rapid growing tree plantations that could reduce a trade deficit in paper products (Bray 1996; Calva Téllez 1997). One sign of the power of community ownership, and of the distrust that rural people feel toward legal schemes that appear to alienate land, was that forest communities largely refused to enter such ventures. At present, then, there is a nascent corporate plantation sector that has had little success in gaining access to land, a more or less disorganized community forestry sector that is dominated by timber poaching and logging contractors, and a growing group of successful community forestry businesses that are supported by the state and international donors and NGOs (Merino 1997; Klooster 2003a; Bray et al. 2003).

Foresters' Careers: From State Employees to Private Entrepreneurs

As we have seen, Mexican foresters have had to endure great career uncertainty as far back as the founding of the forest service, because they had no security of employment. In order to build a successful career, foresters typically had to patch together stretches of government and private employment, they continually had to keep an eye on political patrons in cities and to maintain relationships with colleagues who might one day employ them. In the early years, most foresters went straight into government service after graduating from Chapingo, while the few who worked privately would sell their services to landowners. These private foresters were almost as office-bound as their government counterparts: The heavy burden of paperwork required to comply with regulations meant that they had to delegate tree marking and surveying to untrained subordinates. In the end, both forest service and private foresters were far removed from the forest, moving within the bureaucratic realities of state and provincial capitals. Whether foresters have worked privately or for the state, they are seen by outsiders as members of a recognizable *gremio*[15], a corporate group of identical training and shared outlook (until 1975, all foresters were trained at Chapingo). Since the 1980s, an ascendant *gremio* of biologists and ecologists has been trained at other universities, most especially at the influential National Autonomous University of Mexico (UNAM). This group, known as the *biologos*, has accused forestry professionals of being corrupt and of lacking ecological knowledge. These arguments have gradually reduced the charisma, prestige, and legal authority of the forestry profession. Under pressure from *biologos* and environmentalists, the 1987 LEGEEPA (General Law of Ecological Equilibrium and Protection of the Environment) relaxed the authority of foresters over management plans and required an ecological component that was approved through an institute of ecology (IEE), through which *biologos* could try to slow or block forest management plans. In spite of these changes, foresters retain the key power of writing forest management plans, signing timber transport documents, and supervising which trees are marked for cutting.

Although Mexican foresters today routinely describe present-day forest policy and management as chaotic and disordered, the sheer number of foresters and the detail of modern management plans ensure that forests are more intensively managed than ever before, at least in terms of the detail in documents and logging plans. The descriptions of chaos probably better describe the uncertain predicament of foresters themselves. During the period of the logging concessions, foresters could be reasonably assured of finding work with the government or with an industrial logging company. This changed dramatically with the cancellation of the timber concessions and the repeated economic crises since the 1980s. The state can only employ about 300 foresters, a relatively small proportion of the total number: The rest must find work with logging companies or as private foresters, where they become intermediaries among landowners, logging companies, and the state. Foresters' power and remaining economic opportunities rest on their legal authority to mark timber, sign documents, and write management plans, but this has to be set against the ruthless competition with rival foresters, who may cut prices or launch accusations of bribery and corruption. This is an uncomfortable and uncertain living, marked by doubt, skepticism, and the constant possibility of loss of employment. Faced with such uncertainty, private foresters have little time and interest in enforcing regulations; preferring to concentrate on relationships with landowners and bureaucrats. Put simply, foresters' cognitive landscape is dominated not by forests but by political and personal alliances.

Over the last seventy years, the overwhelming power of the presidency has ensured a pattern of personalistic relationships between senior forestry officials and national politicians, producing repeated policy changes, institutional reorganizations, and forest laws. When a charismatic or well-connected official has been able to forge a strong relationship with a leading politician, he has been able to secure increased attention and budgets for the forest service. Examples of this include the administrations of Enrique Beltrán (1958–1964) who worked during the presidency of Adolfo López Mateos (1958–1964) or the recent administration of Julia Carabias (1994–2000) who worked during the presidency of Ernesto Zedillo (1994–2000). The power of these relationships has dictated frequent reorganizations and a rapid succession of new forest laws (see table 2.1). Increasingly since 1968, each president has sought to address the declining credibility of the state by distancing himself from his predecessor. This also takes place at the level of the forest service, as environment secretaries campaign against past corruption and devise administrative reforms that distinguish the new administration from its predecessor. At times a new government may seek to give the appearance of a "clean sweep" of old and "corrupt" officials by employing a whole generation of young foresters in high positions, as under President Miguel de la Madrid (1982–1988). These frequent reorganizations create a pervading sense of insecurity: Officials may be promoted to the highest level in one administration, only to lose their employment entirely in the following one.

Bureaucratic Transparency and the Opacity of Local Knowledge

Over the last century, forestry regulations, generated and controlled in Mexico City, have declared themselves as instruments of official transparency. Paradoxically, this is a transparency that has made local ecological knowledge of all kinds particularly opaque to the state. This can be illustrated by regulations surrounding agropastoral burning and logging, topics that have been of central concern to the state since the foundation of the forest service. Fire and logging regulations are formally and symbolically controlled by high-level officials in Mexico City, but this has resulted in practices of evasion by foresters and rural people who must conceal their actions from official notice. Highly trained officials and poor indigenous farmers face the same dilemma: They have to exercise their practical and theoretical knowledge of forests and ecosystems, and they must do so in the teeth of a bureaucratic declaration of transparent oversight of the smallest details of what happens in the countryside.

Detailed and unenforceable fire control regulations have persisted over the last seventy years, continually seeking to make burning practices legible to the state and continually having the opposite result. The most recent fire regulation required farmers to present a written application to the forest service containing the time, date, location, and design of the proposed burn (SEMARNAP 1997c). As the forester Javier Mas Porras skeptically pointed out, this could not possibly have been enforced:

> thousands of applications would arrive . . . and could neither be revised nor approved for lack of technical personnel . . . the norm would put more than

ninety percent of *campesinos* in the situation of ecological criminals due to not complying with the numerous restrictions which are imposed upon the authorization of their ancestral practices. (Javier Mas Porras; quoted in Borgo 1998:174)

Almost seventy years after the fire regulations of the 1930s, the forest service had resorted to an almost identical but even more complicated set of regulations, ignoring the forest service's failure to enforce previous regulations. Fires, as the apparent enemy of the forest, are set by problematic and uncontrollable swidden agriculturalists who are the polar opposite of the order and control represented by the forest service. Fire and rural disorder therefore sustain official order: This is a generative pairing that appeals to the imaginations of politicians and urban audiences who support the forest service even as fire regulations continue to be ignored (Mathews 2005).

Official transparency has proved as hostile to foresters' knowledge as that of farmers. The science of silviculture (see box 2.3) was imported from Europe by Quevedo in the early twentieth century through a cadre of French forestry professors hired to teach in Mexico City in 1911–1914 (Junta Central de Bosques Y Arboledas 1912; Departamento de Bosques 1912; Junta Central de Bosques Y Arboledas 1911; Vera 1903; de Lioucourt 1898). Over the last fifty years, silvicultural treatments (i.e., logging) were tightly restricted to selective logging that would preserve tree cover and protect watersheds. This silvicultural system was ecologically unsuited to regenerating pine species, which require larger cuts that produce scarified mineral soil and higher light levels. The main object of regulation was to forbid clear cuts and to enforce a minimum diameter of cut trees, which could easily be verified at road side checkpoints. The administrative needs of the forest service and the relatively poor road system and mountainous terrain made it practical and desirable to control logging systems not in the forest but through checkpoints at key roadblocks.

From 1943 to 1980, foresters were forbidden to use their professional judgment as to the best silvicultural system for any particular area of forest. In practice, of course, they could do whatever they wished, as the forest service was highly unlikely to verify logging practices. As with the case of agricultural burning, the concentration of symbolic power in Mexico City effectively made all local accommodations with ecological or political circumstances potentially illegal. Like peasant farmers and illegal loggers, foresters were forced to engage in evasions and concealments in order to exercise their professional judgment. Foresters bitterly resented this restriction on their technical authority, but they only succeeded in claiming official recognition of their de facto autonomy in 1980.

One way to understand the strange form of authority promoted by the forest service is to consider the temporality of official knowledge. From Quevedo onward, official declarations, management plans, and forms tried to perform the forest service as a solid and stable institution that was the source of order and knowledge and that reached to the farthest reaches of the forest through detailed regulations. In fact, the forest service has been notably unstable, with frequent reorganizations and new forest laws (see figure 2.2 and table 2.1). Management plans have typically run for only five or ten years; in combination with the instability of the forest service, this has almost completely obscured the effect of past logging operations on forests. It took many years before the forest service realized that the mandatory selective logging system was counterproductive because it largely prevented the regeneration of the pine species that the regulation sought to protect (Snook 1986; López 1979; Musalem 2000). Fire control regulations have been similarly double edged. On the one hand, regulations have permitted the forest service to represent itself as the source of order and stability, but on the other hand they have made invisible the beneficial effects of light fires and intentional agropastoral burning. Paradoxically, fire is essential to ensure the regeneration of the pine species that are the main source of commercial timber, so that official environmental discourse of fire as dangerous has actually contradicted local ecological reality and would have completely prevented pine regeneration had it ever actually been applied.

In imposing a single silvicultural system on the forest, the state was imposing a method of seeing forests in the present, and of tending, ordering, and creating new forests for the future. In order for the silvicultural system to be applied successfully, the boundaries of each stand of trees in the forest would have to be mapped and kept track of over the lifetime of a tree, from seedling to final cut. In the case of the pine and pine-oak forests of Mexico, this was around sixty years, but as we have seen, the state's control over forests was tenuous and unstable. Forest management plans moved from "community" forests in the 1930s to private control in the 1950s and then to parastatal control in the late 1970s, but the actual forest land was not owned by the logging companies, and the forests were not mapped out into areas of forests known as "stands," the units that could be managed over the long term. This meant that the official silvicultural system functioned as a rhetorical claim and a gesture toward the future, rather than a continuous system of managing present forests. Until the 1990s, therefore, foresters effectively had to act as if the forest had never been logged before. Even as they imagined the sixty-year cycles that would guide tree seedlings into future trees, foresters had to deal with the six-year cycle of official and political life. To date, the only management plans that I have been able to find that actually reflect on the impact of past exploitations on the present-day forest, were written for forest communities in Oaxaca in the late 1990s (CEMASREN 1999; SARH and UCODEFO #6 1993). This is, to say the least, a dramatically counterintuitive finding. Contrary to official representations of forest communities as being unstable, it is the forest communities rather than the federal bureaucracy or individual foresters that have had the ability to retain records of past forest management and, potentially, to focus on the ecological particularities of the forests they control.

Political Culture: Stable Discourses and Unstable Institutions

From its beginnings in the early twentieth century, the founders of the Mexican forest service enlisted the landscapes of the Valley of Mexico and the scientific theories of desiccation and silviculture in order to claim the authority to preserve forest cover and halt rural burning. Watershed protection and concerns over flooding had been perceived as critically important in the 1920s and 1930s, and even when the science of desiccation fell out of favor in the 1950s, the concern to protect forest cover remained embedded in logging regulations and concepts of agricultural and forest land as strictly separate. The early days of the forest service remain embedded in other ways: Through the continual effort to assert authoritative knowledge of the small-scale details of burning, logging, and timber transport, the state continues to practice a system of documentary transparency that produces official ignorance of what happens in much of the nation's forests. Even as individual regulations are criticized or evaded, the *style* and form of regulations has remained the same: Forms and regulations require ever more information and rigor. The failure of individual regulations should not distract us then from the impressive and continuing success of the assertion that the state reasonably ought to control the minute details of what happens in distant forests. The reasonableness of this project remains more or less unquestioned; this is a cosmology of order that is enduring and stable in the face of shaky institutions and faltering careers.

The uncertain careers of officials and the instability of forestry institutions have persisted alongside the impressive stability of official discourses of reason, science, and technical knowledge, and their opposites, uncontrolled burning, rural ignorance, and disordered forests. The particular political culture of Mexico has translated silvicultural science into a particular kind of authoritative official declaration of knowing and managing forests, even as state institutions have been unable to sustain foresters' practical and scientific knowledge of forests. As silviculture emerged from laboratories and experimental forests and came to Mexico, entrepreneurial technocrats drew on the authority of science to support the authority of the state through a demarcation of politics from science, reason from ignorance. Alongside official knowledge claims, the workings of administrative politics effectively silenced the ecological knowledge of indigenous farmers and of government foresters alike. This may appear to be an indictment of a weak and imperfect Mexican state, a story about corruption in far away places, but there is much evidence to show that this tension between knowledge and authority is not a predicament of defective Mexican forestry institutions but of bureaucracies in general. In many other places and at other times, governments, bureaucracies, and companies that draw on technical knowledge may simultaneously silence, conceal, or distort the knowledge they claim to speak for. What is, however, unique about Mexico is the particular relationship between the state and the indigenous communities that have come to control forests.

In this chapter, I emphasized that the particular contexts of Mexico City, the Valley of Mexico, and the moments of formation of the Mexican forest service affected how globally mobilized knowledge about forests came to be incorporated into projects of state-making and public reason in particular places. A further set of contexts and actors have affected these processes: Forest landscapes were inhabited by indigenous people who were organized into coherent communities with centuries long traditions of agriculture and political action. These were places where the technical and the political were remade. Forests are also actors in this story, as past histories of forest growth produce the landscape that present-day actors try to manage or protect. In what follows, I will describe how a particularly potent set of indigenous forest communities came to become interlocutors for state-sponsored forestry and how a particular landscape has changed, partially in response to human imaginations and desires and partially as a result of the energy and power of living things that live and die in ways that humans cannot entirely predict or control.

3

The Sierra Juárez of Oaxaca: Mobile Landscapes, Political Economy, and the Fires of War

Standing at the crest of the Sierra Juárez, one feels far from cities, governments, and forestry bureaucracies. Range after range of fir and pine-clad mountains recede into the distance (see figure 3.1). The air is cool, the tropical sun stingingly hot. At first this might feel like a forest in Arizona or Spain, but the combination of fir trees, cactuses, and flowering agaves quickly reminds one that this is a tropical mountainous place, not a temperate pine forest. How did these mountain forests become entangled in the projects of the Mexican state? What kinds of resistances did bureaucratic projects of legibility and official knowledge encounter from indigenous forest communities and remote forests? This landscape was not a blank slate on which foresters and officials could write freely when the forest service gradually arrived here in the 1940s. On the contrary, the particular history of this landscape produced the political and environmental actors with which the Mexican state had to deal, the changing forests and the unruly indigenous communities that somehow had to be woven into forms, regulations, and official ways of knowing. The Sierra Juárez is the kind of mountainous pine forest that the Mexican state tried to incorporate into modern economies of knowledge over the last century: The predicament of the indigenous Zapotec and Chinantec people who live here resembles that of the indigenous people who inhabit and own much of the forests of Mexico. Beyond Mexico, the ways that environmental history produced legacies that limited and conditioned state power, official knowledge and popular understandings of the state, can help us understand the predicament of expanding state bureaucracies in general. Despite their best efforts to declare a rupture with the past, all knowledge institutions that seek to control landscapes must encounter and somehow domesticate distant spaces, where social and environmental change have produced particular ways of understanding the state and of making a living on the land.

Figure 3.1
The Sierra Juárez looking north from Benito Juárez.

This feels like a remote place; the cold breeze and the hot sun, the forested mountains, and the sound of the wind in the trees seem far from government offices and forms, from traffic-filled cities and noisy markets. Ecological and political histories show something quite different: a history of connection, disconnection, and reconnection that has linked forests and political economies, indigenous communities, and the nation state. An attentive reading of forest ecology and the structure of present-day forests reveals a past of agriculture and mining, of warfare and conquest, of plagues and resettlement. Looking downhill to the north, fir trees and stubby short needle pines drop steeply away toward the steamy heat of the Valle Nacional; only thirty kilometers away, crops of sugar cane, banana, and tobacco can be grown. Up here, it is always more or less cold; on afternoons in the rainy season, towering cumulus clouds sweep in from the north, and often torrential rains set in and no one can go outside. I well remember setting a forest camp with logging technicians at three o'clock on a summer afternoon because it was too wet to go outside until the following morning. Turning to face southward, toward Ixtlán and the densely settled Valley of Oaxaca, rainfall

Figure 3.2
The Valley of Oaxaca looking south from El Punto.

drops quickly and the less dense seasonally dry pine forests are more hospitable and inviting to human settlement, to farming, and to sitting outside on a summer afternoon (see figure 3.2). This apparently remote place is not far from large cities, dense human settlements, and powerful state bureaucracies.

This chapter describes the history of the landscapes of the Sierra Juárez by thinking through the effects of political economy and state intervention on forests and fields, on indigenous political institutions, and on forms of land ownership. This is a reading that runs largely against the grain of the authors' intents; such a reading of political economy allows us to trace a broad sweep of landscape change over the last centuries, to account for the advance and retreat of forests, the movements of towns and settlements, and the entanglement of particular local species of animals (cochineal insects), vegetables (pines and oaks), and minerals (gold and silver) in global circuits of exchange. However, such accounts of expanding economic systems or state bureaucracies can easily become rigid structures; it would be easy to write an environmental history from which trees and people almost entirely disappear. One way to avoid this is to pay only conditional allegiance to such categories as states, markets, and economies, to treat them as instruments for asking questions about nature, while looking always for the resistances and the unexplained excesses that such questions encounter. A second way to avoid the determining effect of political economies is to take trees seriously as living and growing actors, which move across the landscape on their own account and according to their own time scale, too slowly for us to see unless we use the tools of forest ecology, stand dynamics, and fire ecology. Forest ecology offers another way to encounter the resistance of material things to projects of knowledge; it helps us to come at the history of forests from a direction that is not entirely determined by either political economy or histories of institutions.

States claim to know and act on nature: The Mexican forest service has claimed to manage and protect forests from fire and indigenous people. All of this might seem familiar; around the world, fighting fire seems like an eminently ordinary thing for states to do (Pyne 1998). In this place, fighting fire is actually strange and unnatural: The pine trees that dominate these forests need fire in order to regenerate. Further, histories of fire and warfare have been almost completely suppressed, so that when I came to the Sierra Juárez in 2000, almost no one willingly talked of past fires. Everywhere, there was evidence of fire, and everywhere, there was a kind of structural amnesia: Officials, foresters, community leaders, and young people all agreed that this was a natural place and that fire was an enemy of forests. This chapter therefore pays particular attention to evidence of fires, to agricultural fires in fields, to fires that burn pastures and encourage grass regrowth, to the small fires that can burn through an open forest of pines and leave them only slightly scorched, or to the raging crown fires that burn down entire forests. Fire plays a role in this landscape, but it has largely been silenced and forgotten: Future chapters will look at the reasons for this forgetting, but for now I will outline the multiple lines of evidence for a history of fire. One way to look for past fires is to look for the ability of some kinds of trees to resist and thrive in the face of burning. I used the methods of fire history to look for scars on trees that survived remote fires and grew up to dominate forests or colonize abandoned fields. Such studies, which rely on statistical calculations and the materiality of trees, do not replace or transcend evidence from archives or oral histories. Fire history studies are a way of asking questions, they have their own limitations and resistances: We can date fires to a year, sometimes to winter or spring, but seldom to the timing of the specific droughts or rains that determined whether a field came to be dominated by pines or oak trees. Further, we have little way of knowing why a fire came to be, whether it was set by a farmer, a soldier, or a lightning strike.

Political economies and states are often framed as structures or powerful extra-local actors that intrude on local places, making landscapes and nature the product or prey of these forces and failing to allow for people's ability to use forests in the making of identities and politics. I have tried to avoid this appearance of structural forces imposed on passive local places by combining political economy, politics, and fire ecology in order to think about environmental history in another way, as knowledge of a lively nature produced by multiple instruments of knowing, and as the partial engagement among markets, politics, and species, each moving through time at its own pace and with its own logic, affecting one another and yet never fully determining each other.[1] Political economy, institutional history, and fire ecology are instruments of knowing. Each of these instruments of knowing has its limitations, encounters its own resistances, and leaves some things unexplained; this is the excess that allows us to understand the limits and pragmatics of the knowledge produced by each instrument.

The Environment of the Sierra Juárez: Topography, Climate, Vegetation, and Cultivation

Standing on the crest of the Sierra Juárez, one stands on a division between major climate regimes: Ecosystems and human settlements are dramatically affected by the vast variation in soils and climates across only sixty kilometers. This diversity of ecological zones over such a relatively short distance has had a corresponding impact on human settlements and agricultural practices. The Sierra Juárez is a moderately high subrange of the Sierra Madre Oriental, which separates the steamy tropical lowlands of the Gulf Coast from the drier high valley of Oaxaca to the south (see figures 3.2 and 3.3). These are rounded, forested mountains that seldom rise above 3,000 meters, the peaks are below the tree line, and there is no permanent snow. During the summer months, moist air rises over the Gulf of Mexico in the morning and is pushed toward the Sierra Juárez by the predominant northerly (*norte*) winds; towering cumulus clouds collide with the mountains, and torrential afternoon rains fall on the highest elevations. By the time the clouds then sail over the Valley of Oaxaca, much of their moisture has already fallen. The

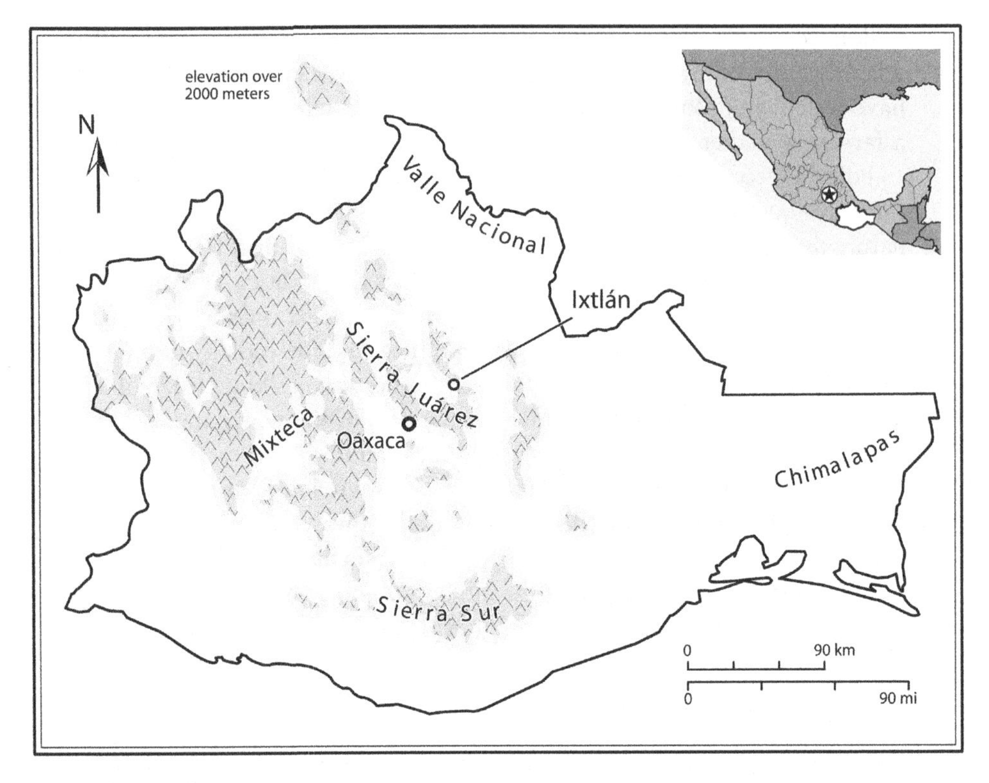

Figure 3.3
State of Oaxaca with mountain ranges and regions. *Source*: Map by Metaglyfix.

valley is in the rain shadow of the mountains, summers here are hot and dry, and the late summer rainy season is relatively short. I have spent many summer afternoons in the city of Oaxaca watching cumulus clouds drift over, remembering how cold and wet it must be in the pine forests on the northern side of the mountains.[2]

In the Valle Nacional on the far side of the mountains, the climate is humid and warm: The original vegetation was tropical lowland forest dominated by evergreen broadleaf trees growing on deep lateritic soils. Beginning in the late nineteenth century, much of this forest was cleared for tropical agro-export crops, such as pineapple, sugar, bananas, and tobacco (Bartra 1996). The middle elevations[3] on the northern side of the Sierra are covered by cloud forest (*bosque mesofilo*); this climate zone is ideally suited to coffee plants, and much of this forest has been cleared for smallholder coffee growing.[4] Higher still, between about 2,200 and 2,800 meters, the forest begins to be dominated by pine species. These tropical pines have long needles that hang down limply; on summer afternoons, they are coated with moisture from fog or rain that nourishes epiphytes on older trees. This is a dense forest of pine trees, broadleaves such as oaks are present, but they rarely dominate, and few human settlements are now present here: It is too cold, too wet, and too difficult to sustain agricultural crops. The rapidly growing "pino rojo" (*Pinus patula*), with long needles and a thin reddish bark, is the most important timber tree, although the slower growing *P. ayacahuite* and *P. pseudostrobus* are also valued. Water is everywhere, from running streams to steadily dripping trees. Fire might seem to be impossible in such a place, but these forests have a history of intense fires that consume large areas of forest; many trees have deep fire scars on their trunks. The dominant pine species are in fact adapted to fire by prolific seed production, rapid growth, and serotinous cones that remain attached to the tree and open only in response to a fire. After fires dense stands of pines rapidly colonize old burns.

At the very peak of the range the climate is hostile, with frequent frosts and occasional snowfalls during the winter. The forest at these elevations is dominated by the fir species, *Abies hickelii* and *Abies oaxacana*, intermixed with slow growing *Pinus rudis*, a short, dark gray barked, stubby needled tree. Dropping down the southern side of the Sierra Juárez toward Ixtlán and eventually toward Oaxaca, the climate becomes rapidly warmer. In the past, arable farming in the cooler temperate zone between 2,000 and 2,600 meters produced wheat, barley, and corn, but since the 1960s, much of this land has been abandoned, and what little agriculture remains is maize cultivated on irrigated fields near villages. Rainfall is lower and more seasonal here, with a long dry season.[5] Late onset of rains can result in almost complete crop failure, and farmers have to time their plantings carefully, making use of irrigation from streams where they can. Traditionally, farmers spread their fields across elevations and used multiple crop varieties as a form of insurance against drought, but now most of these fields have been abandoned. The pine-oak forests at these elevations are more fire prone than the moist forests to the north: The pronounced dry season dries leaf litter, allowing relatively frequent light fires with shorter flames and lower temperatures. Such fires do not usually kill adult trees, and the dominant pine and oak species (*Pinus teocote*, *P. oaxacana*, *P. douglasiana*, and *P. michoacana*) have thick barks that resist fire. Below this zone are drought deciduous forests and deciduous evergreen thorn forests that reach down to the river valley of the Rio Grande, where cochineal cactuses were formerly cultivated. In the Valley of Oaxaca, these forest types have largely been cleared for cultivation, but in intermontane river valleys, such as that of the Rio Grande immediately to the south of Ixtlán, the thorn forest type is still present and expanding to occupy abandoned cattle pastures.[6]

Conquest, Demographic Collapse, and Commodity Production

The most important human and environmental event of the last five hundred years was the devastating impact of the Spanish conquest. When the Spanish arrived in the early sixteenth century, large areas of what is now forest were cultivated, and human settlements and trade networks spread throughout the mountains. Between 1530 and 1600, populations decreased by perhaps 90% due to epidemic disease and warfare. New technologies and new plant and animal species affected how the survivors made a living, as iron tools, wheat, barley, sheep, goats, and cattle came to be gradually incorporated into local livelihoods and ecologies. In the wake of this catastrophe, forests regenerated on abandoned agricultural fields and terraces. Some settlements were abandoned as colonial authorities resettled remnant populations near churches and gradually imposed the colonial political/legal organization of indigenous communities, the *republica de indios*[7] (Chance 1989; Whitmore and Turner 2001).

Since this initial population crash, the Sierra has been repeatedly affected by cycles of commodity booms and busts that have in turn affected settlement patterns, agriculture, and the forests of the area. When human populations partially recovered in the early eighteenth century, so too did agriculture, and forest was cleared for new fields. Expanding indigenous populations provided the labor to support a boom in cochineal dye production in the eighteenth century, and in some communities, people partially abandoned subsistence agriculture in order to specialize in dye production. The cochineal insect, which grows on cactuses cultivated in the warmer semitropical valleys on the Atlantic side of the Sierra Juárez, had come to link indigenous farmers and international trade networks (Romero Frizzi 1988; Hamnett 1971). The wealth provided by the cochineal boom provided the means to build elaborate churches in communities with access to the cochineal cactus growing areas (e.g., the churches in Ixtlán and Guelatao). These churches are the focus of community life to this day, a material trace of political power and international trade. During the wars of independence and reform (1810–1860), cochineal trade networks were severely damaged and production declined precipitously (Romero Frizzi 1988:107–181), only to be finally supplanted by synthetic aniline dyes in the 1850s. A modest boom in silver and gold mining during the late nineteenth and early twentieth centuries shifted trade and forest clearance to mining areas. This natural resource boom, like the cochineal trade, proved to be relatively ephemeral: A crash in silver prices and the effects of the Mexican Revolution (1910–1920) caused the almost complete abandonment of mining. From the 1950s onward, industrial logging of the pine forests in the cooler part of the Sierra Juárez has connected forests to cities, but like previous episodes, this commodity boom was also of relatively short duration. Timber production peaked in the late 1970s, and the forest communities of the Sierra now produce much smaller amounts of timber. Each commodity boom has linked particular ecological zones to world markets and disfavored other areas, causing shifts in population, the clearance of forests, or the abandonment of fields. In the warmer and more humid northern side of the Sierra, coffee production has increased sporadically since the late 1930s (Nader 1990), causing the widespread conversion of cloud and semitropical forest to coffee groves. Here, too, commodities have been fickle and uncertain forces. From the 1980s onward, coffee cultivation has become increasingly less profitable due to declining world coffee prices and competition with cheaper producers in other countries. Significant areas of coffee cultivation have been abandoned, and people have migrated to Mexico City and the United States (González 2001).

Each of these natural resource booms had impacts on the environments and the social organization of the Sierra communities, and each boom added to a fund of experience of interacting with outsiders. This experience ranged from knowledge of how to manipulate bureaucracies—as the rich record of colonial period law suits demonstrates (Guardino 2000; Chance 1989)—to the participation of *serranos* in the civil wars of the nineteenth and early twentieth centuries, to new understandings of the value of pines and oaks for timber production or biodiversity protection at present. The most obvious impact of the environment on people's consciousness is the term *serrano*, which literally means "someone from the mountains." People from the Sierra Juárez have come to share an identity as people of the mountains who have different interests from the inhabitants of the nearby central valleys of Oaxaca. This shared identity has been powerfully invoked when *serranos* enter state and national politics.

The Sierra Juárez as Military Landscape

The indigenous communities of the Sierra experienced considerable economic success during the cochineal boom of the eighteenth century, as well as around regionally important mining centers (see figure 3.4). The collapse of colonial rule and the almost continuous wars of independence and reform (1810–1865) disrupted complex credit and trade relationships, while aniline dyes replaced cochineal dyes by the mid-nineteenth century. Former cochineal farmers had to become farmers or pastoralists, merchants abandoned trade, areas of forest were cleared for agriculture in areas where paid work had sustained food imports, while in other areas declining markets for food caused diminished cultivation and a greater reliance on subsistence agriculture.

By the 1850s, with the exception of small-scale silver mining in the hands of *mestizo*[8] political entrepreneurs, the Sierra communities were relatively disconnected from broader circuits of trade and exchange. This commercial decline affected politically sophisticated indigenous communities with a tradition of well-organized community militias led by powerful *mestizo* military/political leaders. *Serranos* had been involved in international trade and were willing to become important actors in national politics from the mid-nineteenth century onward. Sierra communities supplied organized militias first to the liberal president Benito Juarez (1858–1872) who was born in Guelatao, a few miles from Ixtlán in 1806, and later to his successor, the *caudillo*[9] Porfirio Díaz, who was political and military chief of the district of Ixtlán in the 1850s. *Serrano* troops fought directly under Díaz's command during the wars of Reform (1858–1861) and French Intervention (1862–1867), and during his successful rebellion and ascent to the presidency in 1876 (McNamara 2007), an office that he retained with only a brief interruption from 1876 to 1911.

Díaz's local allies were the mestizo military/political leaders of the *serrano* forces, Francisco Meixuiero and Fidencio Hernández, who organized local militias and manipulated municipal positions within the leading towns. With the relative calm and the capital investment-friendly policies of the later *Porfiriato*,[10] investment in mining and agricultural production for export increased greatly and began to restore connections between the Sierra Juarez and international trade circuits, albeit without significantly affecting local land ownership patterns and community institutions. Meixueiro and Hernández used their political clout to negotiate a limited degree of economic development

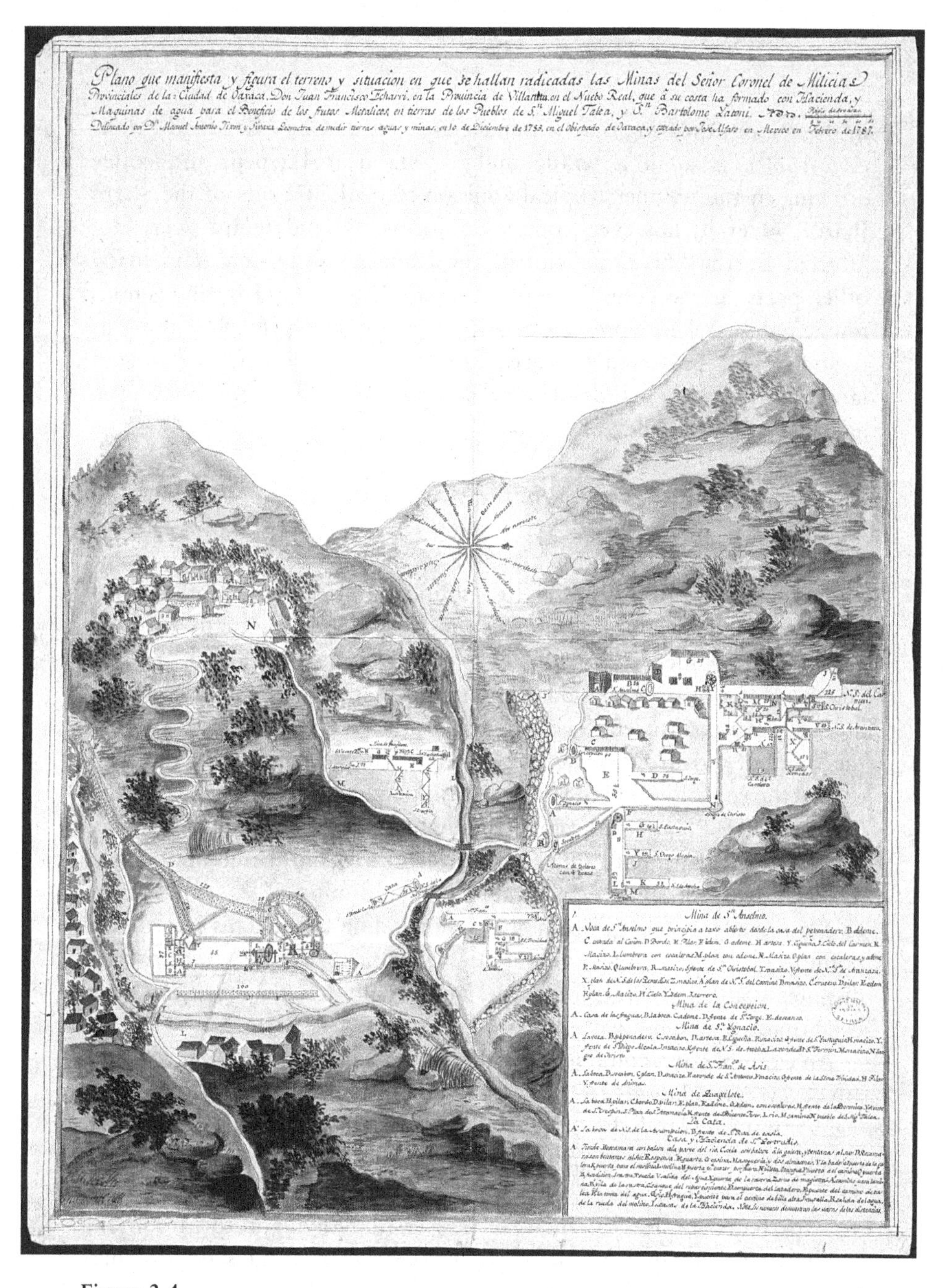

Figure 3.4
Mining landscape near Villa Alta in the the Sierra Juarez, 1785. This kind of highly planned and ordered landscape drew upon indigenous labor and caused the clearance of nearby forests. *Source*: Carlos Weinold, Plano de las Minas de Juan Francisco Ecjarri en la Provincia de Villalta.: España, 1785. Courtesy of Ministerio de Cultura, Archivo General de Indias.

in the region, focusing on the mining area around La Natividad (Mairesse 1880), a small a textile mill at Xía near Ixtepeji, and coffee growing in the warmer tropical zone on the Atlantic side of the Sierra Juárez. Overall, however, communal forms of land tenure were not affected by the liberal policies of the Mexican state, which in many other parts of the country resulted in the loss of land by indigenous municipalities. The close personal connection between Díaz and this region largely protected this area from the dramatic concentration of landholdings in the hands of great landowners, which took place even as nearby as the Valley of Oaxaca.

Making Communal Territory in the Sierra Juárez

The present-day distribution of land in the Sierra Juárez is in some measure a reflection of the relative political and economic influence of indigenous communities during the nineteenth and twentieth centuries. Sierra communities accommodated state projects of mapping and of encouraging private freehold land tenure by mapping only their boundaries. They were sufficiently cohesive and politically powerful to prevent the kind of privatization and fragmentation of communal land that caused the consolidation of land ownership in much of the rest of Mexico (Craib 2004). State projects of mapping and territorialization did, however, offer an opportunity for powerful and old communities such as Ixtlán to form alliances with surveyors and military leaders in order to claim control of large areas of land, including forest. This is a touchy point to this day: Smaller communities often have a history of territorial disputes with their larger neighbors and usually feel that their land has been unjustly taken away.

An example of a community with a long history of legal and political advocacy and solid title to land is Ixtlán de Juarez. Ixtlán had established legal title to its lands by 1722 and was able to make use of the latest topographical science to map its boundaries in the nineteenth century. This was a sign of the presence of a group of educated people in the area, perhaps notaries, teachers, or lawyers. Following in this tradition, by the 1930s, Ixtlán was able to support the career of the schoolteacher, historian, and ethnologist, Rosendo Pérez García, who exhaustively chronicled the history of the Sierra, highlighting the cohesion of communities and the relative eminence of Ixtlán in particular (Pérez García 1996a [1956], 1996b [1956]). By the mid-twentieth century, Zapotec intellectuals like Pérez were beginning to formulate local histories and ethnographies. This is an early example of indigenous people speaking for and about themselves, although for Pérez, indigeneity was of less importance than the celebration of Ixtlán as a source of order and progress, of undoubted community cohesion, and of solid legal title to land.

Writing in 1956, Pérez described a struggle surrounding the mapping of Ixtlán's lands in the late nineteenth century, which caused a brief insurrection by Ixtlán's neighbors (Pérez García1996a [1956]:281–283). The insurrection was pacified by the personal intervention of Porfirio Díaz himself, who acceded to rival communities' demands that a German topographical engineer be expelled. Although this effort to map the Sierra had been partially successful at best, a copy of this engineer's map is preserved in the town hall of Ixtlán, and the modern boundaries of the community forests largely follow this map. Rather than being a successfully imposed project of state legibility, mapping was the product of local political entrepreneurs' opportunistic employment of surveyors. Rather than leading to a systematic cadastral map of community lands, which could have led to the disintegration of community land tenure, Sierra lands were only mapped on a community level,[11] and land was not made available for sale and consolidation into large holdings (Garner 1990, 1988).

The struggle over Ixtlán's boundaries reveals how the military power of the *caudillos* could translate into bureaucratic power with long-term political and environmental consequences. Ixtlán's "scientific" and "modern"[12] map helped solidify its boundaries after the Mexican revolution and assisted in its later battles with mining and logging companies. To this day, control of a complete *carpeta basica* (basic folder) of legal documents is a prerequisite for a forest management plan that can allow legal logging. Military prominence in the nineteenth century has assisted Ixtlán's prominence in forest protection in the early twenty-first century, a testament to the environmental legacies of Zapotec indigenous militias who were willing to fight in the distant wars of the young Mexican nation state.

The Mining Boom of the Nineteenth Century

Mining of silver and gold had been important during the sixteenth century but declined relatively quickly after the conquest. A period of renewed activity in the late eighteenth century gave way to relative abandonment during the wars of the nineteenth century (Pérez García 1996a [1956], 1996b [1956]; Chassen and Martinez 1990). These mines were small by the standards of the great mines of San Luis Potosí or Zacatecas in the north, but even such relatively small mines were important to the local economy.

The late nineteenth century saw a mining boom throughout Oaxaca, with the richest mines located in the Sierra Juárez (Pérez 1996a [1956]: 315–330; Chassen and Martinez 1990). A large number of people participated in mining, whether as full-time miners in the large mines that employed the majority of the adult men in some towns,[13] as part-time miners in small family operations, as illegal miners on abandoned concessions or in streams, or as suppliers of food and materials to the mining centers. Mining was a highly polluting process: Ore refining required mercury and sulfuric acid, which ultimately ended up in rivers and streams, and some of this mercury is likely present in river sediments to this day. Depending on the quality of the ore and the technology used, a large amount of charcoal or firewood was burned in small smelters; timber for pit props required heavy tree cutting near mining centers, while the long-distance mule caravans supplied chemicals, tools, and explosives. In communities without mines of their own, people produced charcoal and timber or grew food for sale in the mining centers.[14] From 1908 onward, a crisis in mining caused by declines in world silver prices shut down all of the mines in the Sierra, except for the richest and largest at La Natividad. Overall, mining probably affected forests near the mines most dramatically; in these areas, repeated cutting of oak trees for firewood and charcoal will have increased the presence of oak species, which can resprout from stumps, reducing or eliminating pine species that rarely resprout in this way.[15] Large mule and donkey trains needed to supply the mining industry required the support of extensive pastures and areas of cultivation, with the consequent use of fire to produce grazing and the clearance of additional land for food production. The cultural and political impacts of the mining boom were of comparable significance: Communities most actively involved in the mining boom during the nineteenth century are currently more likely to have legal title to large areas of land and to be involved in the intensely bureaucratic practices of community forestry. These communities are also notable for having relatively few speakers of indigenous languages, perhaps due to their greater contact with outsiders.

Impacts of Fire and Grazing on Nineteenth-Century Landscapes

Grazing and pastoral fires and heavy tree cutting for firewood and charcoal combined to reduce or eliminate forests near mining centers and along the mule trails (figure 3.4 shows a mining area in 1785). Burning favored grasses and fire-tolerant pines. Fires every five or seven years would have encouraged open pine savannah landscapes, and more frequent burning combined with heavy grazing would have killed tree seedlings entirely and produced grasslands. Given the present-day stigma attached to fire, it is ironic that one of the most famous of Mexican presidents, Benito Juárez (1806–1872), was a former shepherd (Zerecero 1902:12) who in all likelihood burned pastures. Juárez's success in expelling French forces from Mexico in 1867 was one of the rare occasions during the nineteenth century when European military ambitions were decisively defeated on colonial peripheries, making Juárez a statesman of genuine international stature.[16] As Juárez's birthplace, the village of Guelatao a few miles from Ixtlán has become a powerful symbol of nationhood, indigenous identity, and of Mexico's international significance. Mexican politicians and presidents regularly visit Guelatao on Juárez's birthday, March 21st, invoking his power to unify the nation in the name of universal values of tolerance and respect for indigenous people.[17] The association of Juárez and Díaz with this area caused it to be memorialized in paintings, providing evidence of a landscape that was more intensively grazed and burned than at present. In 1887, Mexico's most famous landscape painter, Jose María Velasco, traveled to this area (Altamirano Piolle 1993, 278–294). His paintings from this visit allow us to glimpse what was already a relatively deforested landscape at a time when mining and the mule transport networks had yet to reach their peak (see figure 3.5).

As well as being a professional painter, Velasco was a naturalist and botanical illustrator with a strong commitment to representing reality as accurately as possible. His luminous paintings incorporated his theoretical commitments and scientific training; his detailed depictions allow a trained viewer to recognize plant species or cloud types to this day. As a member of the *Sociedad Mexicana de Historia Natural*, Velasco was aware of contemporary concerns about the impact of deforestation and drainage on the Valley of Mexico (Sociedad Mexicana de Historia Natural 1883; Trabulse 1992). These interests, together with his desire to produce a dramatic painting, could well have caused him to exaggerate the degree of deforestation, but his firm commitment to representing reality as realistically as possible suggests that this was not the case. Further, this relatively open and pastoral landscape tallies with the evidence of increased pasture and burning to support the mining economy of the late nineteenth century. When I visited Guelatao in 2001, I took

Figure 3.5
Landscapes of Guelatao in 1887 and 2001. Photograph: *View of Guelatao* in 2001 by Julian Priest. Painting: *View of Guelatao* by José María Velasco, 1887. *Source:* Collection of the Museo Nacional de Arte, Mexico City. Reproduced by permission of the Instituto Nacional de Bellas Artes y Literatura.

pictures of the same view, as nearly as possible from the same place that Velasco must have set up his easel.

A comparison between Velasco's painting and the recent photograph shows a landscape that is much more forested in 2001 than it was in 1887, probably due to reduced grazing pressure and less frequent fires.[18] Areas around the mining centers near Ixtlán would have been more severely deforested still, and forests that were further from towns were probably relatively intact.

An account of pastoral burning closer to the city of Oaxaca is provided by the eminent Italian botanist Cassiano Conzatti (1862–1951), who was a rough contemporary of Quevedo. Like Quevedo, Conzatti was familiar with desiccation theory and warned that deforestation caused by agropastoral fires would cause droughts and threaten water supplies:

> Soon I will complete twenty-five years living in Oaxaca, and not one time in this quarter century have I missed seeing . . . fire on the hills which surround it. The summits I say and I do not have to correct my judgment, because this is all that is left. The remaining forest cannot burn because it burned in previous years. (Conzatti 1914)

To a scientist like Conzatti, steeped in desiccation theory and official understandings of fire, it was self-evident that these fires were a disaster. In fact, annual fires would have maintained grasslands, and less frequent light fires would have allowed open pine forests to persist. Such shifts in forest cover to grassland or open forest inevitably cause ecological changes, but these changes cannot easily be called degradation. As many scholars have pointed out, environmental degradation is very much in the eye of the beholder and often depends on how that beholder is using the landscape. From the point of view of pastoralists, this may not have been a degraded landscape at all.[19]

Unleashing the Fires of War: Revolution in the Sierra, 1912–1925

During the second half of the nineteenth century, rural municipalities and indigenous communities in many parts of Mexico were forced by the state to privatize their lands, resulting in the creation of vast landholdings in the hands of well-connected Mexican and international investors. The indigenous communities of the Sierra Juárez are in marked contrast with this history (Garner 1988, 1990; Pérez García1996 [1956]a, 1996 [1956] b). The *caudillos* of the Sierra had relatively little interest in claiming direct control of agricultural land (McNamara 2007) as their authority depended on their ability to manipulate traditional communal government structures rather than on becoming major landowners.[20] When Díaz fell from power in 1911, the *serranos* and their military leaders were largely opposed to the successive national regimes, attempting to stay to one side of national events and briefly capturing the state capital in the name of a state sovereignty (*soberanista*) movement that effectively declared a plague on the houses of all other contenders.[21]

On the outbreak of fighting on the national stage in 1912, longstanding tensions between Sierra communities over land and labor conditions led some communities to question the authority of the *caudillos*, erupting into open fighting in the Sierra also. A vicious settling of accounts ensued, both between rival communities and between rival factions within communities: Perhaps the lack of work in the now collapsed mining economy had made tensions still stronger. A harrowing example of these community conflicts is the history of Ixtepeji, which aligned itself with the Maderista cause (see figure 3.6). In 1912, a group of communities led by Ixtepeji attacked and partially burned its neighbor and historic rival, Ixtlán. In retaliation, an alliance of communities, including Ixtlán, combined with federalist forces to burn the town of Ixtepeji as well as its allies (Ibarra 1975:67).[22] The adult men of Ixtepeji were taken prisoner and sent away to forced labor in distant parts of the country. The Sierra is now so quiet and peaceful that war and disaster are hard to imagine, and it was only when I heard about these events from a witness that I came to realize how violent this time had been. In 2001, a 98-year-old *comunero*[23] of Ixtlán, Luis Ramírez García, told me how federalist forces and their local allies had debated whether to execute these prisoners. In the end, he said, and I found this a chilling detail, the *serrano* leaders decided against executing the adult males of Ixtepeji, not on the grounds of humanity, but because of the health hazard posed by their bodies (interview notes, December 11, 2000).

The years 1912–1925 were a terrible time in the Sierra Juarez, as in many other parts of Mexico. People often had to flee their homes and take refuge in the forests and hills, making it impossible to cultivate their fields and causing widespread starvation and disease. The details of the fighting in the Sierra are complex,[24] but the main point for our purpose is that rival armies moved back and forth through the mountains, often killing draft animals, burning many towns, and on occasion using indigenous ecological knowledge of fire as a weapon in war. Ramírez went on to describe deliberate burning of forests when the Ixtepejanos and their allies were once again chased out of the Sierra in 1916:[25]

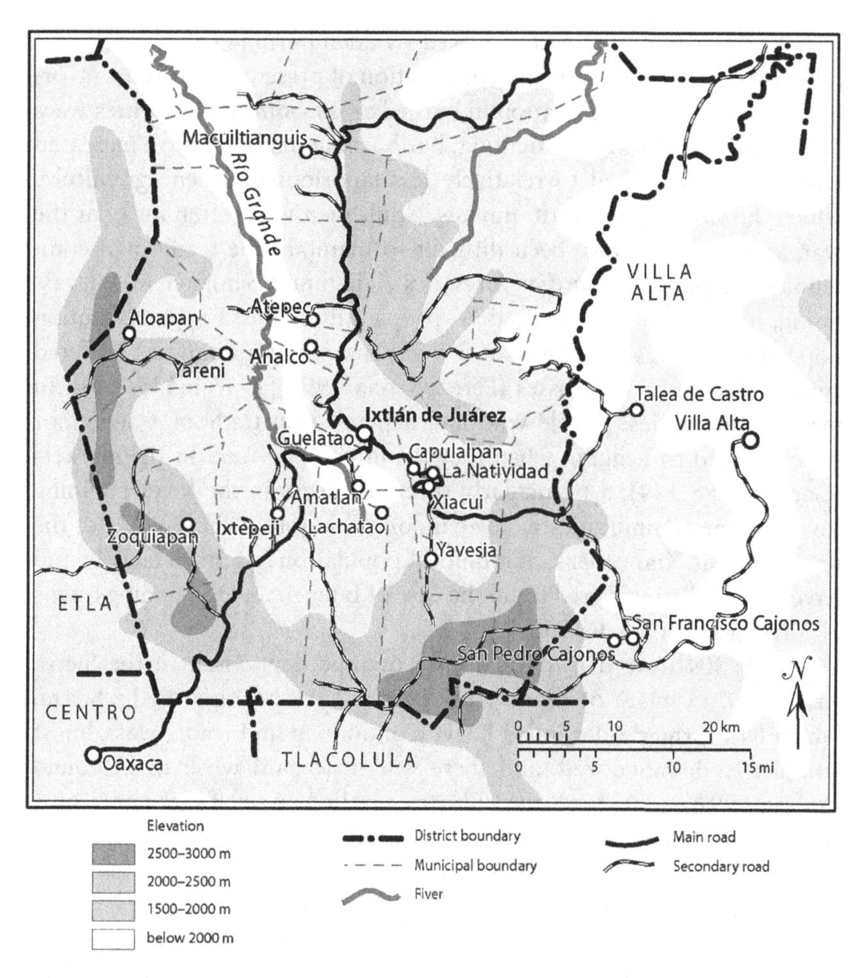

Figure 3.6
Principal communities in the District of Ixtlán. *Source*: Map by Metaglyfix.

They were on a mountain, and we couldn't get them out of there, it was disputed for 8 days. One of the *serranos* thought and at night, at ten at night, he set fire to the foot of the mountain, five or six men with torches. They burned the whole mountain, it was all burning and in that direction the smoke and fire. The fire attacked them and they [the Carrancistas] ran and the *serranos* got in there. (interview notes, December 11, 2000)

The *serrano* general, Isaac Ibarra, describes a similar use of fire in fighting between Los Pozos and La Coronilla in March of 1920, further underlining that fires were a commonly used weapon deployed across the landscape (Ibarra 1975) even if we cannot tell how large or intense these fires were.

Both the fires of war and increased swidden burning have had powerful effects on the structure and composition of present-day forests, favoring increased pine regeneration in burned areas. Some of these fires were deliberately set during the fighting, but burning may also have increased when people switched to relatively less laborious swidden agriculture, which did not require draft animals. Adult men were often away at the war, and it would have been difficult to maintain the traditional community fire patrols (*guardias forestales*). The human impact was devastating; in Ixtlán, Pérez counted 76 people killed out of a prerevolution population of 1,120 adults, and many nearby communities suffered comparable or greater losses (Pérez García 1996 [1956]b:435–446). In many communities, people who had depended on trade or wage labor were reduced to begging when they could not buy food in the markets (Garner 1988:154), a particularly severe problem in the former mining towns. Some communities were abandoned entirely, as in the case of the textile mill at Xía; other communities' populations declined steeply and have never fully recovered (as in the case of Ixtepeji and the mining towns of Yavesia and La Trinidad).

The 1930s were a time of deep economic depression in the Sierra Juárez and in much of Mexico in general. The economy of the Sierra, which had formerly depended heavily on mining and trade, was almost completely de-monetized, and there was little paid work to be found (Young 1982); subsistence agriculture prevailed except for the area near the sole remaining mine at La Natividad (Pérez García 1996[1956] b:435–446). Far from being typical, this period of economic isolation and reliance on subsistence agriculture was quite unusual, marking a pause between the previous cycles of mining, warfare, and political action and a new era of logging in the pine forests and of coffee production on the northern side of the Sierra. Emigration to seek work increased after the United States entered World War II in 1941, as young men went off to find work with the *bracero* migrant worker program; but even as the young men were leaving, the Mexican economy was expanding once again, and roads began to bring trucks and logging companies into the Sierra, creating modest opportunities for paid work. The network of logging roads probably slowed or halted light grass fires, which could have been blocked by the road entirely without any human intervention. Trucks had secondary effects on the environment: Each truck replaced a large number of people and pack animals, reducing the demand for food and pasture for the mules[26] and substituting local agricultural production with imported maize and beans. Cultivated land in areas that sold their produce to the remaining mine at La Trinidad declined precipitously also, a final blow to the local agricultural cash economy. Even as the new roads brought in manufactured goods, government officials, and ideas about forest management, they helped agricultural land return to forest.

Mobile Pines and Changing Forests

So far I have described how the forests of the Sierra Juárez advanced and retreated in response to broad political economic and demographic trends. I have recounted how political institutions structured control of the landscape as indigenous communities adopted colonial legal forms in order to claim territory, and I have described how these factors affected how people used fire in pastoralism, agriculture, or warfare. These ways of thinking about the history of the landscape make forests and landscapes relatively passive, a kind of blank slate on which political economy or warfare is imposed. Forests are more complicated, interesting, and unpredictable than this; state forestry, scientific knowledge, or peasant agricultural practices encounter the uncertainty of lively plants. Pines and oaks continually seek to colonize the interstices of the state and the economy, looking for the possibilities offered by abandoned fields, old burns, or roadside rights of way, proliferating during the particular seasons and moments that favor them. Far from being immobile, then, forests are lively: They continually push back, they expand and move across the landscape.[27]

One way to apprehend the movement of forest across the landscape is by continually imagining the development of forests through time. Foresters do this using the theories of "stand dynamics," which connect the life cycles of individual trees with the development of relatively homogenous groups of trees ("stands") over decades or centuries (Oliver and Larson 1990). In thinking with stand dynamics, foresters try to reconstruct the particular disturbance that gave a tree seedling the opportunity to germinate, flourish, and grow up to become one of the adult trees that now dominates a particular forest stand. Fire in particular is an important disturbance, and pine and pine-oak forest types[28] found in the Sierra Juarez are fire dependent and fire adapted. The resistance of trees to fire, their evolved adaptations to fire, and their ability to take advantage of the disturbances that burning produces, allow fire-adapted pines and oaks to colonize landscapes in patterned but not completely predictable ways.

It has long been known that many pine species are adapted to regenerate and persist under fire regimes of varying intensity and frequency, including, of course, fires set by humans (Agee 1998; Richardson 1998; Richardson et al. 2007). This broad pattern is confirmed by research in Mexican forests, which shows that fire is essential to the successful regeneration of the pines that often dominate the pine and pine-oak forest types (Rodríguez-Trejo 2008). An increasing body of research in Northern Mexico has found complex histories of fire in pine forests fairly similar to the dry side of the Sierra Juárez, where frequent light ground fires at frequencies of three to six years appear typical (Fulé, Villanueva-Díaz, and Ramos-Gómez 2005; Rodríguez-Trejo 2008).[29]

Although the importance of fire to pine regeneration is well known to ecologists, until quite recently it has been systematically ignored and suppressed by the Mexican state, which continues to propagate the myth that fire is an alien and a destructive force. This is paradoxical: Pine trees often owe their presence to past fires, and many of the commercially attractive areas of the forest that are presently being logged are the direct result of a history of burning and fire suppression. Pines are typically pioneer to mid-successional species that are well adapted to regenerate on the abandoned fields and burned areas produced by warfare, agriculture, or forest fires.[30] Because they do not grow well in deep shade, areas of pine forest are usually evidence of some now long forgotten disturbance that allowed a group of pines to regenerate, a pulse of regenerating seedlings that has now become a forest. Contrary to appearances, in relatively well-watered forests, large and small pine trees are usually around the same age; rather than being younger, small trees simply failed to thrive after the initial disturbance that allowed them to regenerate. In mixtures of pine and oak species, pine trees have usually regenerated as the result of one or more past disturbances, such as forest fires or landslides. The oak trees in these forests are usually found growing up beneath the over-story of pine trees, patiently outlasting them and growing rapidly into the canopy when the death of a large pine gives light and space.[31] These broad ecological patterns are highly visible in the Sierra Juárez, where pines dominate disturbed sites such as old fields, burned areas, and power line rights of way (e.g., Negreros and Snook 1984; Snook 1986), and seedlings germinate prolifically on deliberately applied clear cuts (Gomez Cárdenas et al. 1999).

During my numerous visits to the forests of Ixtlán in 2000–2001, I never saw large areas of young pine trees. The exception, which proved the rule, was the few large burn areas in other parts of Sierra, where I

Figure 3.7
Pine regeneration near Yavesia, Sierra Juárez, 2001.

saw dense stands of vividly green young pine trees (see figure 3.7). The lighter area at the center of the photograph is a dense stand of young pine trees that grew up as the result of a large wildfire in 1985.

Contrary to appearances, even very moist forests with long rainy seasons are not exempt from fire (Agee 1998; Goldammer 1993). Even in humid tropical forests that might seem impervious to fire, increased water availability supports rapid plant growth, producing fuel for intense fires when a drought year eventually occurs (Goldammer 1993; Goldammer and Seibert 1990). Pines therefore have a range of fire adaptations, from species in drier areas that resist light fires with thick bark, to species in moister ecosystems that are consumed by rare intense fires but grow prolifically and rapidly on the burned site (Richardson 1998; Richardson et al. 2007). Although oaks are generally less fire resistant, they too often have a thick bark that resists burning and are usually able to resprout from the stump when the above ground part of the tree is killed by a more intense fire. In Mexico and around the world, oaks and pines are very often found together: The timing, intensity, and frequency of fires is one factor that determines whether pines or oaks dominate a forest stand as the result of a particular fire. As in Mexico, pine and oak forests around the world are valued for timber and firewood, and, as in Mexico, typically states and logging companies are interested in pines for timber, and peasants are interested in oaks for firewood, animal fodder, or fertilizer for fields.

The present-day structure of forests is often a clue to histories of disturbances, and forest management plans can be interpreted using scientific theories about the ecologies of species and the development of forest stands. When I looked at the forest management plan of Ixtlán, it was clear that fire had been formerly been widespread, even omnipresent. Across the 8,080 hectares of the commercial pine forests, 4.3% of trees had some kind of basal scar (often the result of a light fire), and 15% of the forest was said to have been "affected by fire" (TIASA 1993:209). The age distribution of the pine forests of Ixtlán, derived from the management plan, suggests that most of the forest stands were fifty to seventy years old, that is, that dominant trees regenerated between 1928 and 1948 (see figure 3.8). The most probable cause for such relatively uniform stand ages is a period of widespread and frequent fires until the early 1930s, followed by a period of fire suppression at least five to ten years

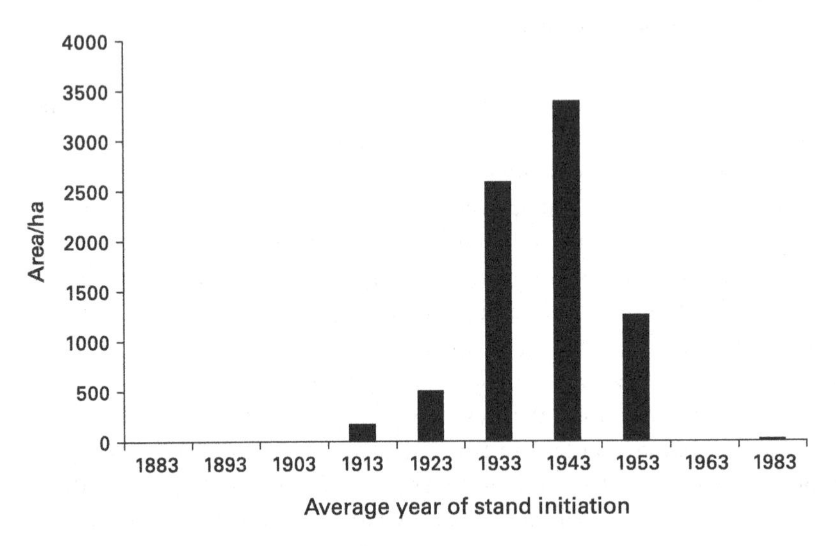

Figure 3.8
Year of stand initiation in the communal pine forests of Ixtlán de Juárez. Ages reported in the Ixtlán management plan (TIASA 1993) agreed closely with my own measurement of the ages of two dominant trees per stand in fourteen stands (Pearson correlation 0.843).

long, which would allow young pine trees to grow tall enough to survive a light fire. Agricultural abandonment alone would be insufficient to account for this age distribution: Regional-level agricultural censuses suggest that at most around 600 hectares of present-day forests were formerly agricultural land.[32] The remaining 7,000 hectares of pine forests are therefore probably the sign of a massive cohort of young pine seedlings that colonized large areas of pasture and open forest from the early 1930s onward, when burning had greatly diminished due to declining agriculture and pastoralism and increasingly effective fire suppression. Pine trees and oaks readily entered the spaces provided by changing agricultural practices, declining human populations, and the absence of fire. This kind of fire suppression may well have been entirely unintentional as even a relatively modest barrier, such as a road or path, can prevent a light fire from advancing during the early stages of fire suppression.

I found a more detailed kind of evidence for histories of burning by making use of the ability of pine trees to resist fires, which however had left characteristic scars at the base of their trunks. A scar may form at the base of a tree when a light ground fire burns the dead leaves and dry debris that often pile up on the up-hill side. This debris pile burns slowly through the bark of the tree, killing a patch of the cambium layer immediately beneath the bark but leaving the tree otherwise healthy (Arno and Sneck 1977). Once such a scar forms, each successive fire is likely to make a new scar so that over the years a tree may have come to acquire multiple scars, recording a series of fires.

The ability of trees to survive fire allows the fire history of forests to be deduced by sampling scarred trees and counting annual growth rings (Arno and Sneck 1977). This is the technical field known as dendrochronology, which is used to date fires and wet and dry years, and, more recently, to look for evidence of climate change. In recent years, such studies have expanded to collect hundreds of tree samples and correlate thousands of scars and tens of thousands of growth rings in order to find fire years across broad regions and to think about the possible impacts of climate change on fire frequency and intensity (e.g., Kitzberger et al. 2007). My own very small study history in two forest stands in the community of Ixtlán sampled only six trees on one site and two trees on another in order to link histories of past fires to the age structures of forests and to gain a general idea of the timing of fires and fire suppression (Mathews 2003). By the standards of contemporary fire history studies, these sample sizes are too small to make claims about "fire years" or "mean fire frequency," but my purposes were quite different. In a context where there was literally no official memory of past fires and burning practices, I wanted to see whether fires had formerly been present and when they had ceased, in order to begin to think about how pine trees might colonize landscapes. When I first began walking through forests with community forestry technicians in 1998, they recounted fire as being entirely unnatural and threatening, and I doubted whether the modest scars at the base of trees could really be due to fires (see figure 3.9). I well remember the first large pine tree that my assistants cut down, my fear that my textbook accounts of fire scars might have led me astray, that I might have killed a tree to no purpose. I was most relieved to see a series of fire scars on that stump, a sign of the ability of this 200-year-old tree to resist fires and carry on growing (see figure 3.10 for a fire scar series).

Working with community members, I cut down trees with scars at the base, removing a "cookie"-shaped cross-section. This section was then polished and analyzed by a painstaking process of counting annual growth rings, allowing me to date a fire by the scar it had formed. The living resistance of the tree to fire, together with statistical analyses and computer programs, the help of my assistants, my ability to get permission from the community of Ixtlán and to negotiate the additional export permits required by the forest service, came together to allow me to construct a 200-year fire history of a small area of forest. This kind of research faces different resistances than does political economy or politics. I had to measure real trees and walk in real forests, I had to pore over tree rings, and I had to perform statistical calculations and produce maps, graphs, and charts. In the end, I concluded that a small area of forest in dry pine-oak forests near Ixtlán had experienced fires about every six years for more than 200 years, until around 1940, when fires disappeared from the scar record (Mathews 2003). Although I could not date fires on the moist side of the Sierra so easily, in this area very hot fires clearly affected the commercially important "red pine" (*P. patula*), as the most recent fires could be dated to the 1950s. Lively trees recorded the scorching heat of fires, and fire scars and trees increasingly opened up conversations about fires in the forest. Once I began to ask questions about specific parts of the forest, about scars on trees, and about what might have caused them, older people in Ixtlán became increasingly willing to share stories of using and fighting fire.

For fire history researchers, correlating fire scars at a reasonable proportion of scarred trees produces a number, an estimate of the frequency

Figure 3.9
Samuel Ruíz's hand near an overgrown fire scar.

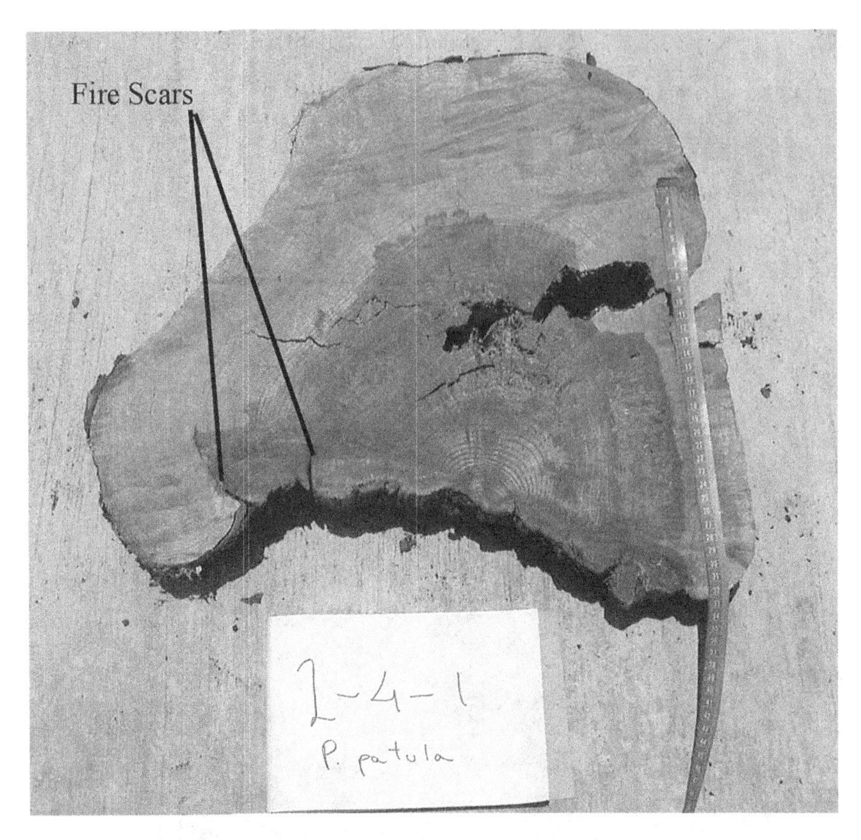

Figure 3.10
Section of a *Pinus patula* showing annual growth rings and fire scars.

with which a light fire passed a given point. Such estimates are hedged with uncertainty, and fire history researchers are careful with such numbers. The ecological impacts of fires are similarly uncertain and critically depend on timing, how long it has been since the last rain, the seed crop on trees that year, the intensity of rains after a fire, the age of the trees that the fire encountered, or how long it had been since the last fire. The time that is made by recording, counting, and statistically correlating a fire scar on a particular pine tree with scars on many other pine trees can only estimate whether a scar was formed during the dormant (i.e., dry) season or the growing (i.e., rainy) season. Fire history then relies on interpretations of ecology, climate, and human history; the resistance of fire scars is both the material sign on which fire history depends and the stubborn limitation of what fire history researchers can know.

Landscapes in Motion: Pines, Oaks, and Fires

A hundred years ago, the botanist Cassiano Conzatti described the annual grass and forest fires in the forests north of the city of Oaxaca. On summer mornings, he would have seen numerous small columns of fire curling up from scattered agricultural fields; he might also have noticed larger fires in pastures, later glowing red into evening. It is striking that there are very few such fires now. Sitting outside on summer evenings in Ixtlán and gazing across a vast sweep of landscape, I rarely saw any burning at all. Large forest fires do occur from time to time, but intentional burning is largely a thing of the past. Reading a landscape for evidence of fire reveals a present absence, the absence of pastoral and agricultural fires that in alliance with humans and stubborn trees have produced the present-day landscape of forests and fields.

Over the last hundred years, the Mexican state and the forest service have been vigorously opposed to all forms of burning, and officials have tried to fight forest fires when they could. In the early twentieth century, Quevedo and his colleagues had justified the authority of the new forest service because they believed preventing forest fires would prevent deforestation and the droughts and floods predicted by desiccation theory. The pine and pine-oak forests that the forest service was to encounter as it expanded into the forests of Mexico were not always places where fire use was increasing; on the contrary, in places like Ixtlán, regimes of relatively frequent burning were to give way to less frequent fires, as agriculture decreased and roads expanded. Present-day forests are the result of these complex histories of burning and fire suppression, agriculture, and agricultural abandonment, but also of forms of political and territorial control, which gave some indigenous communities legal title to large areas of forest. The experience of fire suppression in Mexico echoes experiences around the world, as modern states have intentionally attempted to substitute burning landscapes with the controlled fire of internal combustion in power plants, factories, and cars (Pyne 1993, 1998). In many other places, forest services have similarly misread the landscape, ignoring evidence of past fires and of the human role in producing forested landscapes. Often the state has tried to portray such landscapes as natural and therefore properly under control of the state or of scientists who would know and speak for nature. Environmental anthropologists in particular have mounted a vigorous criticism of such erroneous readings of the landscape, pointing out that what appears most natural may in fact be an anthropogenic forest, a socionatural hybrid of nature and culture (Fairhead and Leach 2000; Kull 2002; Raffles 2002). Sympathetic as I am to such accounts, what I want to highlight here is something slightly different. Rather than reveal that what appears natural is in fact cultural, it is more helpful to notice the continually shifting articulations of multiple nature-cultures, of pine trees with peasants and politicians, of international economies with cochineal fields or silver mines. Pine forests and silver mines are socionatural actors with particular kinds of livelinesses and resistances; they come into history in ways that affect what kinds of natures or cultures they are woven into.

The societies and environments of the Sierra Juárez have a long history of articulation, disconnection, and reconnection with regional and national politics and international political economies, creating a continually moving landscape of fields, forests, and human settlements, accompanied by changing understandings of forests and politics. In some periods, political economic connections affected the environment of the Sierra Juárez extensively and powerfully, as, for example, during the population collapse between 1550 and 1650, and later because of commodity booms in cochineal dye (eighteenth century) and in mining (eighteenth and late nineteenth centuries). Each boom left material and cultural traces: knowledge of trade and debt relations, or of how to organize politically to get land title, expanding areas of pasture, or new settlements and new theories about the environment. At each stage, landscapes and meanings became resources for people who tried to remake their lives. For example, the economic importance of the Sierra communities during the late colonial period was one of the causes for the involvement of *serranos* in the civil wars of the mid-nineteenth century and during the revolution. During the mining boom of the late nineteenth century, the forests of the Sierra were intensely exploited around the more important mining centers, large areas of land were dedicated to food and fodder production in nearby communities, and *serranos* were willing to organize into cohesive militias and fight against outsiders. A return to subsistence agriculture after the revolution and increased burning during and after the revolution gradually gave way to a more forested landscape, as subsistence agriculture gave way to imported food, work in forests, and migration to cities.

A reading of the economic and political history of the Sierra Juárez produces a certain kind of nature, a sense of the broad advance and retreat of forests in the face of waves of increasing population and expanding agriculture, or the advance of forests when populations declined due to colonization, disease, and warfare or changing settlement patterns. By reasoning from the ecological preferences of pines and oaks, I have been able to paint, in broad terms, some of the changing landscapes of the Sierra at a time scale of centuries or decades. More recent political history allows greater detail, with more room for the political logics that drove natural resource extraction, mining, and warfare, and which prompted people to burn forests and towns. All of these, however, are histories in which nature is at best a passive background that suffers the impact of distant forces (for political economy) or local political negotiations and maneuvers (if we think about *caudillos* and their militia followers). In such accounts, the resistance and drama are provided by changing political economies or political affiliations and fail to encounter the resistance of forests or fields. This is clearly not enough: The principal human actors in this history were farmers and pastoralists who would probably recount working in fields and forests as one of the central dramas of their lives. For peasant farmers, lively corn plants and invasive pine trees are passionately interesting and significant.

The liveliness of trees and forests and the skepticism of farmers and pastoralists are not necessarily opposed to accounts of larger political or economic structures. I suggest, on the contrary, that the recalcitrance of fires and fields, of unruly and brave Zapotec militias and of their wily caudillo leaders, are the kinds of fibers out of which the apparently large structures of state or economy are woven. The unruliness of trees and forests reveals the potential instabilities embedded within such accounts of economy and politics. Each way of describing environmental history is a theory that has its constitutive limits; it encounters resistances and absences, which give it its coherence and empirical solidity. These theories are instruments that encounter particular resistances. Political economy encounters the resistance of labor, technology, and power, but it does not contain the unstable performances of politics or the skilled balancing of political relationships carried out by political figures such as *caudillos*. Accounts of political performances and negotiations between military leaders and their *serrano* military supporters encounter the resistance of popular ideologies, political concerns, and ways of making a living, but they have little to say to the skilled work of farmers who burn fields or of trees and plants that grow in these fields. Accounts of fire history try to think with the details of forest ecology but cannot tell us the precise lightning strike or battle that could have caused a burned site to become a pine forest. In each case, the mode of analysis is an instrument that encounters limits and resistances, and the quality of knowledge produced by each instrument can be read in the resistance that this kind of knowledge-making encounters and in the questions that it cannot answer.

As the people of the Sierra Juárez came to encounter the Mexican forest service from the 1940s onward, their own knowledge of farming and forests afforded them a critical distance on state projects of knowing and controlling forests. In a real sense, the resistance of trees and fields imposed a limit on the state's ability to impose official readings of the forest on their understanding of themselves and of the landscape they inhabited. The forests of the Sierra were a material resource for logging, but they also contained legacies of past political events and economies. This history had left a particular kind of forested landscape, inhabited by communities with a history of political connections and relatively solid control over forests, covered by forests that were expanding as agriculture declined and burning decreased. Although the expanding Mexican state largely ignored these histories of fire, political opposition, and local ecological knowledge, these legacies were to become of great importance in environmental politics when Sierra communities encountered the theories of scientific forestry and state-promoted industrial forestry. The ways in which indigenous people understood the state were therefore affected by what they knew about the landscape, by how they made a living in it, and by the long-term effects of past economies, past wars, and past ways of making a living. The stability of state projects of extracting natural resources and controlling forests were affected by people's experiences of stubborn pine trees that came back on old fields or by their memories of agricultural fires that did not go where you wanted them to go. As we shall see, the resistance of the small and the local turned out to affect the stability of state efforts to control the forest.

4

Forestry Comes to Oaxaca: Bureaucrats, Gangsters, and Indigenous Communities, 1926–1956

The pine forests of Oaxaca may appear peaceful, but they are not always so. When forestry science began to travel from Mexico City to the provinces in the 1920s and 1930s, forestry bureaucrats based in the city of Oaxaca encountered a tangled web of political intrigue. Struggles to gain control of valuable timber could become violent, as officials, loggers, and community leaders manipulated forestry regulations and environmental theories to claim control over forests. A distant echo of gunfights, murders, and possible bribery reaches us if we can imagine events in the Sierra Sur of Oaxaca on a winter morning in January 1950. According to newspaper accounts, a survey team from San Juan Mixtepec had been shot at by a group from their neighbor and rival Santo Domingo Ozoletepec, incited and perhaps led by two Spanish loggers, Miguel and José Ranz. The Ranz duo was arrested and sentenced to prison for the shooting: It subsequently emerged that they had paid for a logging road and signed a timber contract with Ozolotepec (Anonymous 1951c, 1951d). The Ranz team sounds like a tough pair: They knew that they had to protect their investment in the logging road and the community political alliance that would give them access to the forests, and they were prepared use violence if they had to. As it happened, there were also less violent means of staying in business. By July 1951, the duo were free, soon after the judge responsible for their release was removed from office for abuse of authority (Anonymous 1951b).

Wildcat loggers such as the Ranz's had to play a number of games successfully: They had to find supporters in the indigenous communities who owned the forests, they had to manage the heavy burden of official paperwork and find official allies in the forest service, and, on occasion, they probably had to buy their way out of trouble. Certainly, the Ranz's did not give up easily: In later years, they were accused of evading tax payment checkpoints (Various 1957), and as late as 1962, they were still in business and still in trouble for ignoring logging regulations (Compañia Forestal de Oaxaca 1962). This was a kind of Wild West in the forest. Logging happened in social worlds pervaded by sporadic violence and doubtful alliances, led by small logging companies that could manipulate or flout the law, paying off judges, avoiding taxes, building roads, and cutting trees. These violent years of wildcat logging, from the late 1930s to the mid-1950s, were to give way to a more orderly situation when two large logging companies gained control of the vast majority of the forests of Oaxaca in 1956. Brief as it was, the period of wildcat logging was a critical moment in the gradual expansion of the Mexican forest service into the forests. Projects of state building gradually entangled the Zapotec and Chinantec people of the Sierra Juárez, introducing them to official theories about forest management and the role of forests in affecting climate. Official projects of fighting forest fires and conceptual divisions between forests and fields gradually transformed indigenous understandings of forests, as previous cycles of warfare and mining became submerged beneath projects of environmental restoration. Even as official theories about the environment traveled to forests, state projects of legibility through forestry regulations failed to find reliable allies, and the Mexican state had little way of finding out what happened in the forest. Critically, the state was trying to know about forests through a fragile web of alliances with loggers, community leaders, and forestry officials: The texture of these alliances made all the difference to the kinds of things that the state could come to know.

Forestry Comes to Oaxaca

The postrevolutionary political settlement had given rise to an increasingly centralizing federal government, which asserted unprecedented control over forests, bypassing state and local governments in the name of science, environmental restoration, and economic development. This program of environmental restoration and political control was built by means of traveling documents and a gradually expanding network of forest service offices where officials lived with the documents, forestry regulations, and forms that were the physical manifestation of the forest laws. The science of forestry traveled to Oaxaca in the hands of a tiny cadre of forestry officials who were responsible for making forests legible and controllable to the state. This was, in the end, an impossible project: There were too few officials, too much forest, and too few roads. Officials did, however, succeed in introducing rural people to official environmental theories, offering them a possible point of entry in their ceaseless search for official allies, whether in disputes over land with their neighbors or in their efforts to find protection from other state institutions.

By the 1950s, national-level officials and politicians considered the model of direct federal regulation a failure, and the Mexican state relinquished direct control over forestry, subcontracting both regulatory and management responsibilities to large private logging companies. Direct bureaucratic control over forests had been defeated by lack of resources and by opposition from regional elites who exploited political contacts and manipulated regulations. Nevertheless, this period was critically important in introducing official environmental theories to a vast range of people all over Mexico, from newspaper-reading urban audiences in Oaxaca, who came to link deforestation with illegal logging and climate change, to swidden agriculturalists who began to realize that their activities were officially frowned on by the state. Inhabitants of forest communities learned that they needed alliances with bureaucratic and financial intermediaries if they were to derive some financial benefit from their forests. Through contact with forestry officials and state rituals of tree planting, indigenous people learned official ideologies of natural resource control: that selective logging protected forests and that forest fires were forbidden and destructive. Rival communities increasingly began to use accusations of clear cutting and illegal burning in order to involve the state in longstanding land disputes. State environmental theories were to be incorporated into popular views of the environment, which form the basis for environmental politics to this day. Official environmental knowledge was a traveling scientific theory that people gradually used in their encounters with trees and fields, eventually remaking their understandings of nature, of who they were, and of what kinds of politics they could engage in.

Cárdenas in Oaxaca: Community Politics and Environmental Restoration

During the 1930s, official efforts to protect forests were accompanied by a renegotiation of the relationship between the state and indigenous communities and municipalities. There were few officials to implement forest laws, so the state had no option but to delegate much of the responsibility for protecting forests to rural municipalities. This can be seen clearly during the visit of President Lázaro Cárdenas to Oaxaca in 1937, where his speeches tried to encourage school teachers to enlist rural people into projects of protecting trees and undermining rural political elites. Cárdenas is a pivotal figure in Mexican twentieth-century history; he is remembered as the founder of the corporatist *Partido Revolucionario Institucional* (PRI) who also confirmed the political settlement that limited a president to a single six-year term, and he is celebrated and revered for nationalizing the Mexican oil industry in 1938. Cárdenas gradually accommodated or neutralized regional political movements across Mexico. Oaxaca, with its notably independent *soberanista* military leaders, required careful handling and a degree of local autonomy.[1] During a presidential progress through Mexico in 1937, Cárdenas traveled to the Sierra Juárez and made a special point of announcing official projects of environmental restoration and political reorganization in Guelatao, symbolically important because of its association with Benito Juárez. In a speech a few days later, Cárdenas made it clear that the Sierra was a peripheral area suitable for projects of environmental protection and natural resource exploitation but hardly a place of political activism or industrial development (Cárdenas 1978[1937]:237–241). Even as he described the Sierra as peripheral and powerless, Cárdenas was engaged in making it so, working to undermine the political authority of *soberanista* leaders who dominated the sierra. Cárdenas encouraged the formation of political interlocutors for the state, fostering PNR[2] committees in rural communities and encouraging secular school teachers and former soldiers to undermine the authority of older office holders, who were allied with the *soberanista* leaders Jiménez and Ibarra.[3] Cárdenas also specifically granted Guelatao full independence from Ixtlán, depriving recalcitrant *Ixtlecos*[4] of the powerful symbolism surrounding the birth place of Benito Juárez. Cárdenas announced state projects of public health and scientific agriculture, founding a hospital and agricultural research station in Ixtlán and a boarding school for indigenous people in Guelatao in 1938.

Cárdenas further affirmed the state policy of weakening the traditional religious-civil community government system by promoting the new institution of the *Comisariado de Bienes Comunales* (Communal Property Commission), separating land tenure from the civic/religious *cargo* system that had hitherto run the sierra communities.[5] The civil-religious system was associated with traditions of fiesta organizing and associated conspicuous consumption; in the anticlerical 1930s, the *cargo* system was seen by the state as a reactionary institution that should be replaced by specifically secular institutions. As we have seen, very little communal land had been alienated during the nineteenth century, so rather than introducing the *ejido* system supported by the 1917 constitution, the new *Comisariado* system provided a new legal form to traditional forms of communal land tenure. Like other state projects, the new *Comisariado* system traveled into the Sierra unevenly, faced by delays, foot dragging, and dissimulation.[6] Even in a "modern" and prominent community such as Ixtlán, the first *Comisariado* was not elected until 1948: Until then there was a completely mixed civic and religious system, and communal church lands were cultivated to provide income for fiestas (Pérez García 1996a [1956]:265–266). In more remote communities, the formal separation of civil and religious institutions must have occurred much later.

The *comisariado* system was to have powerful repercussions for forest exploitation in the Sierra Juárez; it was the *Comisariado* that negotiated timber contracts with logging companies and that was to become the focus of both organized opposition to logging companies and of modern community forest management institutions. The *Comisariado* had a federal interlocutor, the *Instituto de Reforma Agraria* (Agrarian Reform Institute [IRA]), which kept lists of legally registered *comuneros* (commoners) who had the right to elect office holders and to share in income from communal lands. Initially most adult men were inscribed as *comuneros*, but in later years, not all of the sons or grandsons of these men were registered, and in some communities in the Sierra there are large numbers of inhabitants with no legal rights in land. In theory, all meetings of the *Comisariado* were supposed to be attended by an IRA representative who would ensure that a quorum was present, that elections were valid, and that meetings were running smoothly. Thus, the *Comisariado* system allowed a much greater degree of possible federal intervention and official tutelage, weakening the links between the regional power holders and municipal governments.

In his speech in Oaxaca a few days after his visit to the Sierra Juárez, Cárdenas criticized the *soberanista* movement and associated it with the clergy, wasteful fiestas, and oppressive communal labor traditions. Traditional elites caused an unfortunate distance between the revolutionary state and rural people to the benefit of "parasitic elements" within communities, he claimed. Significantly, the official program of secularization was accompanied by a program of environmental restoration:

> The Forestry Department has been ordered to proceed to establish nurseries in various parts of the state, of fruit trees, and to reforest, cooperating with the local government in this matter; it is opportune to insist upon this, before the whole people of Oaxaca, so that they should concern themselves with the con-

servation of their forests, taking care to halt the immoderate cutting which has rendered sterile large regions which were fertile before. [All of our citizens] and especially teachers and pupils should take part in the reforestation of parks, avenues and highways, planting at the same time the largest [possible] number of fruit trees. It is necessary to create awareness of the importance that the tree has in the life of villages. To take care of the forests and to teach children to plant many trees, this should be our motto to enrich the soil of Mexico. (Cárdenas 1978[1937]:237–241)

This was a call for a new relationship between society and nature, linking natural and social order, political and environmental restoration. Just as the state rested its legitimacy on the restoration of order after the moral decay of the Porfiriato and the chaos of the revolution, Cárdenas sought to restore nature after past destruction. He advocated a program to restore the "sterile regions" created by past "immoderate cutting," arguing for a regeneration of both society and nature. The state should, he went on, eliminate "the divisions that for various motives distance the pueblos from the state"; undermining *soberanista* leaders and bringing the state "closer" to the people would inspire civic renewal and moral and natural regeneration. It was secular school teachers who would teach their pupils these new civic obligations though tree planting: The state would control excessive tree cutting in collaboration with communities.

The development projects heralded by Cárdenas were quite different from previous ones and marked a decisive change in how the state was imagining the Sierra Juarez. During the nineteenth-century mining boom, the Sierra Juárez had been seen as a focus for industrialization and mineral extraction, if on a modest scale. State development programs from the 1930s onwards aimed to relieve rural poverty and isolation and conserve the environment. These programs classified one of the most densely populated and politically active parts of Oaxaca as a remote, backward, and definitively peripheral area, appropriate for improved agriculture and forestry, justifying relatively low levels of state investment in a politically problematic region.

Zapotec Views of Forests and Agriculture Before the Arrival of Industrial Logging

The particular environmental history of the Sierra Juárez had produced a landscape that was, by the 1930s, slowly returning to forest in many areas. This history also produced particular understandings of forests: The ways in which Zapotec people understood forests affected how they understood forestry and logging and how they accepted official theories about forests, fires, and floods. *Serranos* took these understandings with them when they went to work as loggers in the forests and forged a hybrid of traditional and modern environmental theories. It was in terms of this hybrid system of knowledge that the abuses of industrial logging concessions were defined in the 1980s, and it is in terms of this hybrid system of knowledge that the legitimacy of the community logging business of Ixtlán is defended at present.

Precisely because the introduction of scientific forestry has had such an overwhelming effect on local people's understandings of the forests, it is impossible to reconstruct from present-day interviews what kinds of theories *serrano* people brought to their encounter with the forest service and the logging companies in the 1930s. At almost the same moment that regimes of environmental control were expanding into the Sierra, state projects of controlling, protecting, and modernizing indigenous people indirectly led to the first wave of ethnographic writing by Mexican anthropologists. Julio de La Fuente's classic ethnography, *Yalalag: Una Villa Zapoteca Serrana* (de la Fuente 1949), describes Sierra Zapotec understandings of agriculture, forests, and climate before the arrival of the logging companies and gives some sense of the impact of official environmental science on local beliefs and practices.

Relationship Between Forests and Climate

De La Fuente tells us that ceremonies to ask for rain were already a dim memory when he carried out his fieldwork in 1937–1941 (de la Fuente 1949:256–266, 303–308). He found that people had a vague belief that witches could ask for rain by carrying out ceremonies near springs high in the mountains and near carved stones associated with pre-Christian "idols."[7] High and forested places were inhabited by the supernatural lord or lords of the mountains (*dueños del cerro*), who could be associated with good crops, rain, witchcraft, or good luck. Springs (*manantiales, ojos de agua*) and waterfalls were both appropriate places to ask for rain because they connected with places inside or above the earth. Human actions such as charcoal burning could anger the earth, and ceremonies had to be carried out in the mountains[8] to placate supernatural beings. Human society was conceptually separate from but materially connected with a powerful nature that might punish transgressions. Punishments could be avoided by rituals that placated natural forces through the consumption of food and drink and perhaps also by sacrifices of animals.[9]

The meticulous detail of de la Fuente's ethnography allows a comparison of the kinds of reasons people gave for droughts in the 1930s with the kinds of reasons they give now. It is highly significant that although people told de La Fuente that the rains were not as regular as formerly, they did not attribute this to the effects of logging or deforestation. Possibly they believed that nature spirits and their management were more important than secular events such as logging. By comparison, at present in Ixtlán, everyone except elders told me that deforestation could cause climate change by reducing rainfall. They were less sure that rainfall declines had actually taken place in the Sierra Juárez, but they were deeply concerned that springs (*manantiales*) might dry up if nearby forests were logged.

At present, for most people in Ixtlán, rain-making rituals are a distant memory, but in interviews with four elders in the community of Ixtlán in 2008, I heard consistent accounts of former rain-making rituals. They told me that until the late 1940s, when rains failed, believers from the community would lead a pilgrimage from the town church to an area of springs called *Los Pozuelos* (little wells), near a small lagoon on a mountainside far above the town. Doña Pérez told me:

> They would go out from the church singing, praying, [to the *Pozuelos*, to the lagoon]. They would do their ritual, that is where they asked for water, I don't know with what prayers the ritual was done, but they would come back singing, and half way down the drizzle would start, by the time they were back in town the water [rain] had arrived. (interview notes, August 6, 2008)

When I asked Doña Pérez why people thought climate was changing, her reply summarized the divergent views of community members and suggested the effect of outside ideologies on understandings of nature:

> It depends upon whom you ask. We *ancianos* (elders) say it is because of the will of god. Younger people say it is because trees have been cut, others say it is because of the ozone layer. (interview notes, August 6, 2008) [10]

Doña Pérez also suggested that climate change was caused by the religious and moral failings of the younger generation, coming close to a view of nature as powerful and affected by human moral choices and ritual practices, rather than the nature degraded by industrial logging described by working age people.

During the first half of the twentieth century, international scientific opinion and forest service ideology broadly agreed that deforestation reduced rainfall and dried up springs; these theories came to Oaxaca with forestry officials from the 1930s onward. Thus, when state foresters tried to limit forest clearance because of its supposedly negative effects on climate, they were arguing for a relationship between forests and climate that may have been more easily accepted by *serranos* because it accorded with their own association of forests and mountains with water. At present, *serranos* more or less universally acknowledge that deforestation causes climate change by reducing rainfall, and they are always deeply concerned that springs may dry up if forests are logged near them. This concern was not reported by de la Fuente, so it is probable that it is an appropriation of state desiccation theory by the communities of the Sierra Juárez, in which internationally circulating environmental theories came to find new allies.

The Impact of the Forest Service Between 1930 and 1956

Forestry officials in Oaxaca faced a conundrum: On paper they had enormous authority to go with their responsibility for forests, but in practice there were too few officials, there were too many communities, and communications were too poor. Most officials had to spend much of their time in town, and rather than enforcing regulations, they delegated responsibility for fighting fires or protecting forests to municipal authorities. Faced with these difficulties, officials tried to inculcate a new culture of protecting and loving trees through the national political ritual of the *Día del Árbol* (Tree Day). This was a performance of official knowledge that sought to convince forest dwellers to protect their forests. The direct impacts of these performances on local uses of forests were minor, but in combination with the modest efforts at direct regulation of logging, they were effective in communicating official environmental theories and models of civic behavior to a broad audience.

In considering a state organization such as the federal forest service, it is easy to confuse the rhetoric and aspirations of regulations with actual influence. Once we begin to look at how many people were involved in applying the regulations, it becomes apparent that regulatory activities are concentrated in space, as around cities, and in time, as at festivals such as the *Día del Árbol.* Although we typically think of the *state* as enforcing regulations, in practice it was usually only one or more state *agencies* that enforced regulations. Thus, when we talk of the enforcement of forestry regulations, we are really talking about the activities of a small number of people (perhaps twenty to thirty for the whole state of Oaxaca in the 1930s[11]) who can invoke the coercive power of other state institutions but must usually move around by themselves, operating by persuasion and education.[12] There was in practice very little coordination between the federal forest service and other state agencies, with the exception of occasional assistance from the army to fight unusually large forest fires. Faced with poor road networks and lacking manpower and vehicles, the forest service only gradually expanded its activities into the Sierra Juárez.

A former head of the forest service, Manuel Hinojosa Ortiz, stated categorically that officials lived mainly in state capitals, lacked rural personnel, and had little contact with forests (Hinojosa Ortiz 1958:32). Ironically, the forest service, which aimed to control the remotest parts of nature, was primarily an urban-based organization and had great difficulty even in reaching the forest areas for which it was responsible. It should come as no surprise that the official ideology of the forest service was heavily biased toward urban views of the countryside, and that urban conceptions of rural life continue to dominate official environmental ideology. The primary point of contact between the forest service and society was through the community authorities, through police control of the city of Oaxaca and other urban centers, and through the annual *Día del Árbol* celebrations. Rather than actually affecting what people did in the forest then, forestry regulations affected intercommunity politics and community members' understandings of state forestry ideologies. The state was very good at communicating what it wanted people to do even as it was very rarely able to make them actually obey regulations.

The federal forest service installed its first office in Oaxaca in 1922 (Anonymous 1922a). Initially, it seems to have concerned itself mainly with trying to control the sale of fuelwood and charcoal in the markets of Oaxaca, affecting people from communities that had supplied the city with fuel from time immemorial. In 1924, we see the nearby town of Cuilapam de Guerrero being censured for charcoal burning (Anonymous 1922b). By 1929, the communities of San Felipe del Agua and San Pablo Etla, also near Oaxaca, begin to appear in repeated conflicts with the forest service and with each other. Both communities produced firewood and charcoal for sale in Oaxaca, which brought them to the attention of the forestry department, while the communities were engaged in a boundary dispute with one another, in which they tried to embroil forestry officials. A series of mutual denunciations show community representatives using official environmental theories in an effort to entangle officials in a boundary dispute with their neighbor.

In a letter to the head of the forest service in Oaxaca, the authorities of San Felipe accused Etla of "immoderate cutting and charcoal burning, which the *vecinos* of San Pablo Etla are carrying out, this is the reason the springs dried up, and affects not only our community but the whole state" (Anonymous 1929). San Felipe's leaders clearly knew the received wisdom about the effect of tree cutting on the springs that supplied water to the city of Oaxaca. Naturally, the leaders of San Pablo Etla denied these accusations, countering them in a letter in which they displayed their knowledge of officially promoted logging practices: "We are not cutting immoderately, but by thinning as you [the forestry department] tell us to" (Anonymous 1929). People already knew that tree cutting was supposed to cause springs to dry up, and they hoped to get the forest service to intervene in property disputes (during the course of the dispute, both sides asked for an armed escort to protect their wood cutters). The desiccation theory espoused by powerful metropolitan environmental scientists like Miguel Angel de Quevedo had clearly already traveled to the leaders of communities near Oaxaca. It is not clear whether people really believed this theory, but it would certainly have accorded with traditional theories about the importance of hills and forests for the production of rain. These arguments also reveal an ongoing relationship between forestry officials and wood cutters, where forest service officials were telling the wood cutters to thin rather than to clear cut. The forest service's demand that firewood cutters should thin would have had little effect on forest regeneration: Most of the trees being cut were oak species, which resprout prolifically from stumps (coppicing). The main advantage of thinning over clear cutting was therefore to preserve the *appearance* of forest by retaining at least some trees standing after firewood cutters had passed through.

By 1930, the forest department was attempting to regulate firewood cutting by requiring communities to form themselves into legally constituted cooperatives (Anonymous 1931a) because of fears about "immoderate cutting and fires on the banks of streams." San Felipe responded by writing to the governor to complain about abuses by forest service officials (Anonymous 1931c), accusing a forest inspector of imposing fines unless they embarked on the expensive procedure of forming a forest cooperative. To add insult to injury, they claimed that when they had invited the inspector to verify that they were logging carefully, he had demanded a large sum of money (twenty-five pesos) to pay for the two-day inspection (Anonymous 1932). This episode is redolent of petty oppression, of the power of documents, perhaps of the inspector's attempts to extract bribes, and certainly of the community's ability to fight back by appealing to the governor. Community petitions were prepared in elaborate copperplate writing on special (and expensive) stamped official notepaper; such petitions required knowledge of the state, of official environmental theories, and of the proper literary style. In any case, there were other ways of getting to market: In 2001, an old firewood cutter from San Felipe del Agua told me how firewood sellers would take circuitous routes into town to avoid the forestry inspector. This was a complex game of urban cat and mouse, played by forestry inspectors and firewood sellers.

Several similar cases concerning communities bordering on the valley of Oaxaca show that this was a typical experience (Lopez Cortes 1930). Although the forest service was quite small, other government officials, or people masquerading as such, could use their authority to extract bribes. In 1931, the head of the forest service in Oaxaca wrote to the chief of police telling him that "various people stop wood[cutters], charging fines as employees [of the forest service]" and that there was no regulation against transporting firewood to town (Anonymous 1931b). Was this a defense of firewood cutters by a principled official or a more pragmatic defense of a revenue source from the depredations of a rival institution? In any case, it is clear that various kinds of officials might ask firewood sellers for their documents, and most firewood sellers would have preferred a quiet payment to a debate over the rights and wrongs of the law.

Outside of the Valley of Oaxaca, forestry inspectors could not realistically police firewood cutting. Nevertheless, forestry regulations were beginning to penetrate the Sierra Juárez, especially because communities that wanted to bring firewood or charcoal to market needed to know how to deal with forestry inspectors on the outskirts of towns. By 1930, the president of Santiago Laxopa, in the very heart of the Sierra Juárez, was writing to ask for a copy of forest regulations "because we lack them" (Lopez 1930), while the associated communities of Lachatao, Yavesia, and Amatlan applied for permission to form a cooperative so that they could sell their firewood in the city of Oaxaca (Garcia Toledo 1930). The small amounts of cash that could be obtained by selling firewood were enough to justify long journeys to the market towns of the Valley of Oaxaca, and firewood sellers were finding that they had to deal with the forestry inspectors who controlled access to markets.

Fire Suppression and Climate Control in the Sierra

Official theories about forest fires were also communicated very effectively during these years. For the forest service, fires were a symbol of disorder and a visible threat to the forests. Fire could not only destroy forests and threaten water supplies, it was also used to convert forest to agricultural land, troubling official landscape classifications. Given the tiny number of forestry officials, there was no way for the state to actively fight fires itself, and forestry officials instead required municipal presidents to organize *Corporaciones de Defensa Contra Incendios* (Community Fire Defense Corporations) (Gomez 1930). The 1930 fire control regulation was extraordinarily detailed and quite impossible to enforce: among many other improbable conditions, communities were required to put signs up around permitted burn areas and ask for written permission from the forest service before carrying out any burn (Gomez 1930; Gutiérrez 1930). Copies of the regulations were sent to all of the municipal presidents in the state (Mares 1932; Dirección Forestal y de Caza y Pesca 1932), but it is likely that these regulations traveled no further than the city hall. As we shall see in chapter 5, a fire control regulation traveled no further in 2001.

How did the communities respond to government attempts to regulate burning given that fires were an integral component of traditional agricultural and pastoral techniques? In 2001, people in Ixtlán could not recall forest service fire regulations as far back as the 1930s, but they claimed that fire fighting was and had always been an entirely community-initiated activity. In general, both officials and *serranos* told me that communities preferred to settle fire problems internally, rather than call in the forest service, which was liable to impose heavy fines.

The state archives contain many cases where one community accuses another of setting fires, either deliberately or through carelessness. Because fire was an integral part of agricultural cultivation, agricultural encroachment on a community's land by its neighbors necessarily implied the use of fire, and many accusations of fire setting were therefore also property disputes. Like farmers, pastoralists and charcoal burners also used fire, and they too were often accused of irresponsible burning. Accusations of deliberate burning were a good way of luring the forest service in property disputes with the neighbors; tellingly, *none* of the accusations of illegal burning is by one private individual against another, *all* are by one community against another.[13] A typical example is the case of San Pedro Teococuilco, which was accused of burning forests on its boundaries with San Juan Guelache, San Miguel, and San Gabriel Etla. The forestry department sent a letter to the municipal authorities of Teococuilco, telling them to cease burning because they were not supposed to convert forest to agricultural uses, as this would "surely damage the springs in this region" (Lopez Cortes 1930).

Accusations about burning demonstrate that rural people knew how to express conventional sentiments about forests and that they could make tactical use of official ideology: It is harder to tell to what extent people actually fought forest fires. In 2000–2001, community members of Ixtlán told me that they had always fought fires, and they could remember notable fires from the 1940s and 1950s. Such accounts of community firefighting were of course partly an effort to describe their community in a positive light. In Ixtlán in particular, official ideology that all fires are bad has made *comuneros*' claim that they fight fires a claim to moral rectitude. However, there is independent evidence that the community of Ixtlán did in fact fight large-scale fires, as in 1942, when a fire on the boundaries between Ixtlán and Atepec caused the authorities of Ixtlán to appeal to the governor for help, accusing the authorities of Atepec of carelessness and indifference (Anonymous 1942). The governor helpfully responded by ordering the communities themselves to deal with the fire and ordering all men over the age of fifteen to turn out and fight the fire. The authorities of Ixtlán probably knew very well that neither the state government nor the forest service had the manpower to help them fight the fire, and they were probably more concerned with strengthening land claims along a disputed boundary with Atepec (Velasco Pérez 1987).

Where the communities chose not to fight fires, forestry officials were impotent; in a 1949 newspaper editorial, an anonymous official described fires in the tropical zone of coastal Oaxaca, blaming deforestation on swidden agriculture and peasant irrationality. He ended on a tragicomic note of frustration: "The fires which are registered in the forests are the exclusive responsibility of the municipal authorities who in no way collaborate in their extinction, and in investigations always blame one another" (Anonymous 1949a). One can well imagine a bewildering cloud of accusation and counteraccusation. While fire remained necessary for farming, the duty to avoid small-scale burning would have remained a conventional sentiment. However, it is likely that communities suppressed fires when they chose to do so, as the traditional use of fire in agriculture was carefully controlled. For urban audiences, official understandings of fire were not challenged by daily work practices, and city dwellers learned to understand fires through newspaper accounts that described uncontrolled forest fires set by irrational and lazy indigenous farmers:

> Everywhere a pitiless destruction of forests is carried out, which it is urgent to radically avoid. Ignorance is in many cases the principal motive for this vicious practice, because enormous fires are produced as a result of carelessness in applying *rozas*. (Anonymous 1947)

State Nature Rituals: The *Día del Árbol*

Logging regulations were evaded by wildcat loggers, prohibitions of agricultural burning by recalcitrant farmers, and firewood cutting controls by wily firewood sellers. All of these efforts were undermined by lack of manpower and by the relative failure of the state to persuade its rural audiences that they should follow regulations and accept official versions of reality. This was clearly well understood by officials, including Quevedo, who attempted to address these limitations by instituting the *Día del Árbol*, a tree planting ceremony that united politicians, municipal presidents, and citizens across Mexico (see figure 4.1). Planting trees and loving trees were to be promoted as civic duties. On this day, the duty of the state and of citizens to protect trees was affirmed in speeches, poetry, music, and tree planting. This was a highly successful event: To this day, the *Día del Árbol* is observed throughout Mexico. Rural schoolteachers, already entrusted with the secularization drive of the federal government, were also supposed to organize students to run tree nurseries: Every year toward the beginning of March, school children distributed trees from the school nurseries.

A program of events from a celebration of the *Día del Árbol* in Comaltepec, in the Sierra Juárez in 1942, evokes a day of marching, loyal speech-making, of standing dutifully in the audience for long speeches. Community officials and school children met at the school tree nursery in the morning: Children spent the day distributing fruit trees through the town. In the evening, the main celebration began with music from the town band, followed by speeches from the municipal president and the school director, an essay on "the tree" read out by a school child, followed by songs, more speeches, and, as the obligatory conclusion, a final band performance (Comaltepec 1942). The flavor of the conventional sentiments and high-flown rhetoric expressed at these events can be captured at another *Día del Árbol* celebration beneath the giant tree

Figure 4.1
Tree planting on the *Día del Árbol*, Xochimilco, near Mexico City. *Source:* Courtesy of México Forestal, *Journal of the Sociedad Forestal Mexicana*, founded by Miguel Angel de Quevedo.

of Santa Maria del Tule, near Oaxaca in 1949 (see figure 4.2). The politician Jose Meixueiro

> pronounced an ardent panegyric of the tree, asking in general terms for protection of trees, for which it is necessary . . . that the municipal authorities and teachers cooperate in the work of vigilance and education, and that they impose penalties upon those who attempt against our forest wealth. (Anonymous 1949b)

The purpose of the *Día del Árbol* celebration was to convince community officials and teachers to enforce forestry regulations and to protect a shared national wealth. These were secular demonstrations of allegiance, one among the many state rituals where citizens and community leaders were expected to listen to official speeches and affirm their loyalty to the state. These rituals performed an ideal world where the responsibilities of state and citizen were clearly defined, where the state was performed as knowledgeable and beneficent, and where citizens affirmed political allegiance, enjoyed the music, and perhaps benefitted from a day off work.[14] Good citizens were supposed to protect the forests, and the state was supposed to help them do this: The speeches and poetry affirmed these principles and visibly demonstrated corporate allegiance to the state. This ritual may have done political work, but it did not connect people, trees, and local landscapes very well. An anonymous newspaper story from 1950 reports:

Figure 4.2
The tree of Tule, near Oaxaca. *Source:* Picture courtesy of Dawn Robinson, 2004.

the sectors who organize tree planting during the week of the tree agree to ask the [department of] agriculture that in the future this activity be transferred to the month of May. The trees planted, apart from the improbable case of being watered, do not prosper for lack of rain. (Anonymous 1950)

As a national ritual, the *Día del Árbol* is celebrated on the same day throughout the country regardless of local ecological conditions. Until 1959, this date was in March (InfoLatina 1998), the height of the dry season in southern Mexico (it now takes place in July). People who attended *Día del Árbol* celebrations must have been well aware that most of the trees being planted were unlikely to survive. This is not to say that the tree planting was solely symbolic: People were appreciative of fruit trees, and almost every town in Mexico is rich with public parks and shade trees. Nevertheless, the mismatch between official planting dates and the rainy season suggests that the political entanglements produced at this event failed to engage with the lives of real trees. Political centralization required officials and their audiences to collaborate in ignoring the uncomfortable and largely unmentionable reality of dying trees. Public knowledge of official tree planting produced the related public secret that many trees were dying.

Logging in the Sierra, 1934–1956

We have already seen something of the confusing and sometimes violent world of the small logging companies and community forestry businesses, which were supposed to be regulated by government foresters and forestry police based in Oaxaca. Initially, the main commercial tree cutting activities were firewood and charcoal production near Oaxaca and timber cutting for ties on the railway line to Mexico City.[15] Logging for pine timber began in the Sierra Juárez in the early 1940s. One impetus for the arrival of logging in Oaxaca was the logging ban in the state of Michoacán in 1945, when a number of displaced logging companies shifted their operations to Oaxaca (Calva Tellez et al. 1989:442).

The bureaucratic practices through which forestry officials tried to regulate logging have remained remarkably similar from 1935 to the present (Calva Tellez et al. 1989). At least until the 1970s, logging permits were approved by forest service offices in Mexico City. Logging companies were controlled by transport documentation that accompanied timber from the moment it was cut in the forest until the moment it arrived at the saw mill or market. Anyone who wanted to cut timber had to pay for a technical plan that described the area of forest and the volume of timber to be cut. This plan would then be forwarded to the forest service in Mexico City, which would issue a book of forms that were filled in with the volume of timber on each truck load and which kept a running count of the total volume produced. Signed documents accompanied logs and personal signatures marked trees through the use of numbered marking hammer, which imprinted the authority of the forester who had authorized the tree to be cut. Forestry technicians and foresters were issued numbered marking hammers and submitted monthly reports to forest service offices in Mexico City (from 1935). An old forester, looking back on forestry practices in Oaxaca in the 1950s, told me:

> The company paid the forester, who was responsible for each area of forest. Many of the foresters wanted to have an easy life, they would lend their hammer to a worker, and allow any kind of cutting to go on. The forester would not even go to the forests. There was one case of a forester who was the owner of a logging company, who had thirty six properties under his management . . . they would

give the marking hammer to anyone. (Epifanio Sanchez, author interview notes, June 1, 2008)

The marking hammer was the material symbol of foresters' authority and professional knowledge. Contemporary descriptions of foresters' work practices agree with Sanchez's account and especially describe a lack of agreement between official documents and forests:

> For none of the ten properties [in Oaxaca] in legal exploitation, can plans be found which fix, even approximately, the forested area, and in some cases, there is no general plan of the property, but only a simple sketch without orientation. . . . Not only the forested area, but even the total area varies according to different studies. . . . (Hinojosa Ortíz 1958:105)

The main method of preventing illegal logging was to verify that documents had been properly filled in and especially that logs were larger than the minimum legal size. The mountainous terrain created natural chokepoints where forest guards, army, or police could set up checkpoints to inspect timber transport documents. People in Ixtlán told me that the system was often abused; even when documents were in order, many checkpoints were not manned by forest service employees, and low-level officials of all kinds were known to demand bribes. Moreover, truck drivers themselves often did not understand the intricacies of documentation and were liable to pay bribes so as to avoid trouble from the police.

During the 1930s, official policy was to encourage the formation of community forestry businesses (Anonymous 1931a; Boyer 2007). Some communities, under pressure from the government, may have formed such cooperatives in order to carry on their firewood and charcoal production businesses. At this time, private logging companies often also incorporated themselves as community logging businesses, as communities did not have the capital, technology, or bureaucratic skills required to log forests legally. Contemporary newspaper accounts describe the means by which outside companies secured permission to log community lands as being pervaded with violence, bribery, or corruption (Anonymous 1959, 1958b). This is the kind of violent environment that the Ranz family was working in. Logging companies had to form alliances with a faction within the community that could deliver a solid coalition in the *Asamblea Comunal* (Community Assembly), which could authorize logging contracts. The reputation of the small logging companies from this period is of violence and fraud: Contemporary newspaper accounts reflect this, with a stereotypical use of the term *rapamontes* (forest pillager) to describe all loggers (Anonymous 1953a).

Who were these logging companies, and how were they able to operate with such relative impunity? Records are patchy, but it is clear that many were politically well connected. For example, members of the prominent Meixueiro and Hernandez families, which had led *serrano* forces during the revolution, were involved in logging in the 1950s (Meixueiro 1954; Bradomin 1953b). These were well-connected regional political elites, in the political language of the day; they were, in the words of one critic, "spoiled sons of the official family" (Bradomín 1953c), generals, former federal or state deputies, and even former forestry officials (Bradomín 1953a, 1953b, 1953c). A typical example of the kind of political and economic career that allowed powerful intermediaries to gain access to forest resources is that of the logger and political entrepreneur Adulfo C. Tamayo, previously a federal forestry official and local deputy, who moved between private employment and a government job.

Biological Impacts of Logging

In the middle of this dizzying array of bureaucratic performances and accusations of illegal logging, it is important to consider what was actually happening in the forests and what impact this might have had on the landscape. Industrial logging focused almost exclusively on pine species, selectively removing large trees and leaving the forest relatively undisturbed. This would probably have increased the presence of oak species; pines require high light levels and disturbed mineral soil for successful regeneration, so selective logging of large pines would have favored shade-tolerant oak species without seriously damaging the forest. The total volumes of timber being logged commercially during the 1940s and 1950s were not large by present-day standards: For example, in 1954, around 93,000 cubic meters were officially cut (Anonymous 1952), compared with more than one million cubic meters per year in the 1970s. Rampant timber smuggling and illegal logging mean that this is very much a low estimate, but the overall impact of industrial logging was clearly less than that of traditional fuelwood cutting, which amounted to perhaps 1,100,000 cubic meters per year at this time.[16] Traditional management of Mexican forests by firewood cutters has largely been ignored and remains more or less invisible to the state, but rural people had managed their forests for fuelwood for centuries, which suggests that firewood management systems could have been fairly sustainable. Most firewood comes from oak species, which can be repeatedly cut, as they resprout from the stump after cutting (coppicing). Coppice management has been carried out for millennia in Europe, and there is no reason to think that the firewood cutters of the Sierra Juárez were less skilled than their European counterparts.

The Defeat of State Elites and the Rise of Industrial Logging

In 1951, the governor of Oaxaca, Manuel Mayoral Heredía, attempted to crack down on the logging companies, summoning them to sign an agreement that they would obey regulations and pay their taxes (Anonymous 1951c). This effort to claim control of logging and its profits was bitterly opposed by state political elites. In the end, Mayoral had to ask for the personal support of the president of Mexico, Miguel Alemán (Anonymous 1951a). Although Mayoral was forced to resign in 1952, federal officials were on notice about the chaotic situation of Oaxaca's forests. Not long after, the federal government decided to promote a policy of awarding large long-term timber concessions to privately owned *Unidades Industriales Forestales* (UIF), cutting state political elites out of the action entirely. Existing logging companies affiliated with state politicians had their logging permits canceled: They had proved too difficult to control and did not have the capital to fulfill the government's plans for industrial development and road construction. The new, much larger timber concessionaires, FAPATUX and *Compañia Forestal de Oaxaca*, were created in part to solve the problems of tax evasion and violence encountered in the 1940s and 1950s. These companies did not necessarily pay taxes on time, but they did submit documents on time, and they were able to maintain a continuous presence in the forests, something that had been beyond the forest service or the small logging companies.

Traveling Theories: Official Ideology and Popular Understandings of Forests

Theories about forests and bureaucratic practices of controlling tree cutting had an uneasy journey to make, first from Mexico City to Oaxaca, and then from Oaxaca to the dispersed indigenous communities that owned the forest. In the end, the state was too thin and too dispersed, foresters were too few, and the forest service probably had little or no direct impact on the agricultural and forestry practices of forest communities in the central Sierra Juárez before the arrival of the FAPATUX logging company in 1956. Official forms, regulations, and public rituals did, however, teach many rural people what it was that the state wished them to do, and that they could use state ideology in order to embroil officials in intercommunity conflicts. Forest fires, as an apparent agent of the destruction of the forest, were portrayed by the forest service as the agent of deforestation, and hence of climate change. Official anti-fire campaigns taught rural people that the state saw burning as morally and legally reprehensible. Community leaders rapidly began to use accusations of irresponsible burning in order to involve the state in conflicts over land ownership: Official environmental science was a language that they could use in order to build relationships with one state institution in order to play it against another, a way of linking their forests with external allies.

The state interpretation of fire as uniformly destructive obfuscated the dynamic relationship between forests and agriculture. As we have seen, forests in the Sierra Juárez have repeatedly been cleared for agriculture and then abandoned to regenerate to forest. In forest communities, many areas of young forest were known as the property of the person who had originally cleared the forest, who had last farmed them, and who might wish to farm there again; for the forest service, these young stands of trees were forest land in need of protection. Officials were only too aware that agriculture was expanding at the cost of forests, but they did not note the opposite transition of agricultural land returning to forest as old fields were abandoned and colonized by young trees. This failure of official attention was partly because of the general advance of the agricultural frontier in many parts of Mexico, but it was also due to the conceptual separation of forest from agriculture, which obscured the dynamic relationship between the two.

In communities nearer to the markets of the Valley of Oaxaca, firewood sellers learned to engage in various legal and illegal strategies of complicity and evasion. Sometimes they tried to bribe forestry officials and police, and at other times they circumvented checkpoints entirely, always affirming that firewood and charcoal had been obtained legally. Firewood and charcoal sellers had to learn officially approved logging techniques and environmental theories in order to defend themselves from the authorities. Because of the official obsession with permanent tree cover and a firm and permanent boundary between forest and field, the forest service had promoted selective logging and the cutting of dead wood, practices that left forests looking relatively intact (Various 1958). Firewood cutters soon learned a prophylactic language where they would claim that the wood they were taking to market was "dead wood" or had been cut by "thinning" a forest stand, rather than by clear cutting. This was a legal fiction, as people admitted to me in 2001–2002. In the end, then, forestry science and official theories about fire or climate change were introduced as prophylactic discourses; it seems unlikely that they had much effect on how people understood forests. A second effect, much less obvious to officials, was how the chaotic and dangerous years of wildcat logging had performed the state. Rural people and urban people alike had learned to look with suspicion on state-sponsored projects of environmental restoration. With the arrival of the logging company FAPATUX in 1956, the scale and organization of logging changed dramatically. Rather than being transmitted through regulations and bureaucratic performances, official theories about forests were inculcated through labor organization and tree marking. Indirectly, the road network that allowed trucks to extract timber caused the abandonment of agriculture and the gradual increase of forests, but this was invisible to the state. Young pine trees colonizing abandoned fields were invisible to the state, as were shoots growing on the stumps of oak trees cut for firewood. Expanding forests, living trees, and mobile agricultural fields were rendered invisible by official efforts to produce environmental order.

5

Industrial Forestry, Watershed Control, and the Rise of Community Forestry, 1956–2001

In the end, the Mexican state lost its patience with the small logging companies. It had had enough of violent conflicts and tax evasion by wildcat loggers, enough of the disorderly habits of rural people who burned forests and degraded the environment and who lied or evaded visiting officials. In 1956, the right to log the vast majority of Oaxaca's forests was awarded to two large timber companies, *Fabricas Papeleras de Tuxtepec* (FAPATUX) and *Compañía Forestal de Oaxaca*, extinguising the logging rights of the small logging companies and effectively subcontracting the enforcement of logging regulations from the federal forest service (Presidencia de la República 1958a, 1958b). The official justification for awarding control of the forests of the Sierra to the FAPATUX paper company is described in a revealing memorandum to the Secretary of Agriculture:

> [Except for a few inaccessible areas] all of the rest [of the forests] have suffered more or less intense exploitation, and all have suffered the impact of indigenous demographic pressure. Large areas have been deforested to dedicate them to a rudimentary and migrant agriculture; forest fires are intense and frequent, destroying the young trees and therefore the regeneration capacity of the forest. Uncontrolled grazing by goats and sheep completes this picture of enemies of the forest. It is no novelty but a great truth, to say that forests subjected to a technically directed exploitation, improve their silvicultural conditions and increase their productive capacity. . . . (FAPATUX 1956)

The FAPATUX logging concession, which allowed the company to exploit the forest, was justified by the degrading agricultural practices of poor, shiftless, and ignorant indigenous people, who burned forests for swidden agriculture and to encourage grass for their goats. This was a nightmare of immiseration: Population growth was placing ever greater pressure on the environment. This attack on traditional agriculture and pastoralism was no more than a restatement of the longstanding official discourse about *milpa* agriculture and pastoralism. Such talk had been repeated endlessly since the 1930s, as the Mexican state attempted to stabilize the conceptual boundary between forests and agricultural fields by sedentarizing supposedly "migrant" swidden farmers and pastoralists who might transform forests into fields. Nevertheless, there was something new here: The eradication of swidden agriculture and pastoral burning and the introduction of a rational science of silviculture were to be organized and paid for by logging companies themselves. Projects of scientific forestry and of governing forests would travel not with a handful of forest service employees who had so far dramatically failed to control the forests of Oaxaca, but with hundreds of company employees, the technicians, foresters, road builders, truck drivers, and loggers who would work and live in the pine forests.

Let us jump forward for a moment to the present, when the accepted culprits of environmental order and destruction have become quite different. The indigenous communities demonized by FAPATUX and the federal government have succeeded in claiming control of their forests. Far from being the authors of environmental degradation, they have succeeded in blaming FAPATUX and, to a lesser extent, the federal government for environmental degradation. Far from using fire in agriculture or farming, community members now see fire as an alien and destructive force in the forest, and memories about the former uses of fire have been almost completely suppressed. This is a dramatic change indeed. In this chapter, we will trace how official ideologies of fire fighting and cognitive categories of field and forest have come to be internalized by rural indigenous people. If the period of wildcat logging can be cartoonishly described as a period of prophylactic commitment to official categories and rules about forests, the period of the timber concessions is when scientific theories about forests came to affect actual work in the forest. Working in the forests in turn affected how indigenous people understood both forests and their relationships with the state. This was an intensely practical and material process; changing labor relations and logging technologies, changing landscapes, and changing forms of communication have allowed remote indigenous communities to impose on government officials an entirely new definition of the world. Far from being a story about the oppressive imposition of an ideology of state-sponsored forestry, this is a story about politically active indigenous communities becoming epistemic actors who remade collective knowledge about the history of forests and the role of the state.

The FAPATUX letter specifically highlighted the isolation and distance that, it was claimed, caused indigenous agriculturalists and pastoralists to destroy forests. Development and environmental restoration would be produced by strengthening connections to the broader economy, through road building, cutting firebreaks, mapping forests, and paying forest guards. This discourse of indigenous isolation was politically self-serving and in many ways incorrect. As we have seen, *serranos*' relative economic isolation was a recent event, rather than being the nomadic subsistence agriculturalists imagined by the state, many people in the sierra had long specialized in small-scale commercial agriculture, while their political activism was only a recently solved problem. This official letter then omitted the extralocal factors of politics, warfare, trade, and mineral exploitation, which had linked the Sierra to a wider political economy and to national political events. Such an account might have found quite different culprits for environmental degradation. In many ways, the official account of environmental degradation has completely changed: By 2000, it was not the remoteness of the sierra that was seen as the cause of environmental degradation, but the closeness of the state, of corrupt logging companies, of dangerous roads. This reversal allows us to look for something else, to ask how and why "remote" indigenous communities have been forced by apparently powerful outside institutions, such as the forest service and the state government, to change their version of reality. How did this new collective, this new public fact, come to be in the world, and with what material, political, and moral consequences?

In 1983, FAPATUX was expelled from the Sierra Juárez by an alliance of forest communities that attracted national attention for their cause. This movement was the product of a new hybrid political identity of *serranos* as careful managers of the forest; subsequently, the forest communities of the Sierra have attracted international attention as exponents of community forestry and sustainable forest management. FAPATUX loggers and foresters used traveling scientific theories about forests and bureaucratic forms of regulation in their encounters with trees, logging roads, and timber transport documents. These practices also produced new environmental identities that could be turned against the logging company in order to form alliances within the state. Burning and environmental degradation came to be known very differently, and community members came to know themselves differently also. A new political identity as virtuous fire fighters transformed *serranos*' understandings of the forests and of themselves, displacing memories of traditional fire use.

Community members learned to see forests through the theories and practices of industrial forestry, marking, cutting, and measuring timber and territorializing the forest into a system of logging roads and management units. This was not a case of the imposition of a state project of knowledge or of the self-imposition of a governmentalizing discourse by rural people. Traveling theories had to be applied to lively natures; these theories never fully explained *how* they were to be applied, nor could they predict how living trees, muddy roads, and stubborn machinery would respond. This resistance then called for skilled practical knowledge; resistance was a resource that allowed *serranos* to form new political subjectivities, new understandings of nature, and new understandings of the state.

Work in the forests under the direction of FAPATUX foresters profoundly affected people's identities, as measured by their adoption of official fire discourse and the practices of industrial logging and fire fighting.[1] How much room for maneuver did individuals and communities have to modify this discourse? It is clear that FAPATUX's ability to affect the conditions of work in the forests was a critical factor in imposing official theories on rural people, but this was not a one-sided business. There was considerable contention about where and how trees were to be cut and how the profits from logging were to be divided; in addition, the experience of working in the forests produced political identities that formed the basis for community opposition to the logging company. Community political power is most visibly manifested in the way that communities have forced the state to shift degradation discourse from blaming indigenous people's use of fire and swidden agriculture to blaming the predatory logging practices of FAPATUX and state-sponsored logging companies. The communities have been able to produce a representation of environmental degradation that forest service officials have come to accept; the relative power of organized forest communities and the desire of the forest service to maintain timber production have created a discursive political playing field where knowledge of forests can be stabilized.[2]

In contrast to the minuscule staff of the forest service, FAPATUX employed hundreds of permanent employees and thousands of part-time workers. Work in the forests under the direction of FAPATUX foresters and technicians gave *serranos* intense exposure to the theories and practices of industrial logging, causing profound and enduring transformations in local environmental politics, conceptions of nature and culture, livelihood practices, and memories of environmental change. FAPATUX deployed the same official environmental ideology as the forest service, but it could communicate it much more effectively, through both the increased frequency of contact with community members and its direct control over the practices of forest exploitation. Official environmental theories were also propounded by the *Comisión del Papaloapan*, an integrated watershed management agency that was supposed to prevent floods through forest protection and agricultural extension activities, and that was active in the Sierra between 1946 and 1982. A still more important impact of the *Comisión* was the road network it built beginning in the early 1950s. These roads caused the displacement of local commercial agriculture by imported food and allowed the penetration of FAPATUX logging trucks into the remotest regions of the Sierra. The decline in local agriculture and pastoralism and their associated uses of fire further assisted the dominance of official environmental theories about fire and forests. Commercial forestry came to substitute small-scale commercial agriculture as a source of income for communities near the mining centers, and the system of knowledge surrounding *milpa* cultivation and fire use became increasingly marginalized within the Sierra communities.

The final government agency to penetrate the Sierra Juárez was the *Instituto Nacional Indigenista* (INI; National Indigenous Institute), which was founded in 1949 but had little presence in the Sierra until the 1970s. The INI was responsible for assimilating indigenous communities into mainstream Mexican society (Aguirre Beltran 1979:143–151), mainly by coordinating the development activities of other government institutions. For the communities of the Sierra Juárez, INI's presence seems to have been perceived primarily as an office in the city of Oaxaca to which *comuneros* and community leaders could go in search of support in dealings with other government institutions. People in Ixtlán did not often talk about indigenous identity, and a specifically indigenous political identity seems to have penetrated relatively shallowly, as community identity is more important and more often invoked. Nevertheless, indigenousness as the opposite of Spanish speaking, education, and modernity are well-entrenched categories that have the potential to become politically salient. Education in Ixtlán is in Spanish only, and few people in the area speak Zapotec.[3]

The Floods of 1944 and the Creation of the Papaloapan Commission

In September 1944, heavy rains unleashed devastating floods on the lower watershed of the Papaloapan, causing flooding in Veracruz and the adjacent lowland portions of northern Oaxaca (Tamayo 1982[1954]; Pérez García 1996a [1956]). In the Sierra Juárez, there was little loss of human life, but the floods washed out bridges and roads and gravely damaged crops. President Miguel Alemán toured the affected areas: A subsequent study by the National Irrigation Commission (Tamayo 1982[1954]) called for a program of dam and road building and the control of erosion and deforestation in the Papaloapan watershed.[4] The institution responsible for carrying out this ambitious program was the *Comisión del Papaloapan*, which disposed of large amounts of money for watershed protection projects, inaugurating a new regime of environmental control in the Sierra Juárez (see box 5.1). The floods of 1944

Box 5.1
The Papaloapan Commission, Big Dams, Big Dreams, and Big Disasters

The commission's principal activity was to build large dams of dubious value in the valley of Tuxtepec: the Temazcal or Miguel Alemán dam in 1949–1954 and the Cerro de Oro dam in 1974–1989. Both of these dams were supposed to provide flood control and irrigation water. The flood control benefits were elusive, while the dams flooded tens of thousands of hectares of good agricultural land and caused the forced resettlement of Mazatec and Chinantec indigenous people by the "hydraulic police" of the Commission. In the case of the Cerro de Oro dam, more than 22,000 people were moved, sometimes more than 250 kilometers away, supposedly with resettlement packages that included running water, electricity, modern housing, and adequate land, but these promises were not kept (Bartolome and Barabas 1990).

The career of Jorge L. Tamayo, who was head of both the Commission and FAPATUX from 1975 to 1979, vividly condenses the importance of personal connections for the weaving of political and economic power. Tamayo was related by marriage to President Jose López Portillo (1976–1982) and was responsible for the pharaonic Uxpanapa project of 1974–1979, which resettled indigenous swidden agriculturalists into newly cleared areas of tropical moist forest with the goal of making modern permanent rice farmers. Not surprisingly, it was a catastrophic failure, and most of them fled the resettlement area. The combined arrogance and fantasy of the Papaloapan Commission are perfectly summarized by its motto "Seamos realistas: hagamos lo imposible" (Let us be realists: let us do the impossible) (Bartolome and Barabas 1990:117). For a review of the health effects of dam projects around the world, see Lerer and Scudder (1999). The failures of the Papaloapan Commission are classic examples of the high modernist failures described by James Scott in *Seeing Like a State* (Scott 1998), showing a similar disregard for local ecologies and points of view.

were a pivotal event: For the first time, the modern Mexican state was explicitly connecting agricultural practices and environmental degradation in the forests of the Sierra with the safety and development of distant urban and industrial centers. Further, the state was prepared to spend money to implement projects of environmental control.[5]

An immediate effect of the flood was the decision by the federal government to ban all logging in the Papaloapan watershed in 1949 on the grounds that deforestation could cause flooding (FAPATUX 1956:26). Small private and "community" logging companies were shut down, leaving in operation only the timber companies that had the political connections to protect themselves. This ban on logging explicitly connected environmental protection to control of natural resources, an ideological move that the communities of the Sierra Juárez took note of in their own future efforts to take control of the forests. The *Comisión* was never as aggressive in the Sierra Juárez as it was in the nearby Valle Nacional, where catastrophic dam building and forcible resettlement projects destroyed the livelihoods of tens of thousands of indigenous people (Bartolome and Barabas 1990). Although this kind of forced resettlement never happened in the Sierra Juárez, people in the area must have heard of what was happening. This would undoubtedly have further strengthened a general awareness that state-sponsored environmental projects and distant environmental concerns could be linked to their forests by the right (or wrong) kinds of alliances.

A contemporary newspaper account of the *Comisión*'s activities describes both its projects and the environmental theories that justified them. It is striking how closely the *Comisión*'s rhetoric resembles the language deployed by FAPATUX at the same time:

> Although in the upper zone of the Papaloapan there are large forests of timber, most of its area is deforested because of the systematic burning of areas dedicated to crops, areas which for the most part are used for one or two years until erosion finishes the soil and leaves rock uncovered. The farmer is not interested in this, accustomed as he is to choosing a new place to begin anew burning, planting and harvesting. (El Imparcial 1956)

The *Comisión* combated the supposedly destructive effects of traditional agriculture by road building, rural electrification, and promoting the planting of coffee and fruit trees, all with the aim of sedentarizing the supposedly "nomadic" peoples of the Sierra Juárez. In the Sierra Juárez, most communities have been settled since the colonial period, and one wonders how the "nomadic" agriculturalists of official imagination were supposed to have produced the ornate baroque churches around which many communities are organized. Clearly, the language of statebuilding required images of rural disorder and mobility that were demonstrably at odds with the most cursory glance at the landscape.

The *Comisión* had a long career, ultimately disappearing in the 1984, but during most of its existence, it had much larger resources at its disposal than did FAPATUX. In any case, to many people in the Sierra Juárez, the distinction between the commission and FAPATUX was unclear. The two worked hand in glove with one another, especially between 1974 and 1979, when the geographer, Jorge L. Tamayo, was head of both FAPATUX and the *Comisión*. This blurring of two state environmental bureaucracies makes good practical sense: Without the road building of the commission, the logging company could not have existed, and the two shared a common environmental ideology. Although the *Comisión* had a regional office in Ixtlán, it has left little imprint on popular memory, and its road building activities are remembered as having been carried out by the logging and paper company with whom people interacted on a daily basis.

Industrial Forestry Arrives in Oaxaca

The logging concessions covered enormous areas of forest and contemplated an entirely new intensity of territorial control. FAPATUX was awarded a logging concession over 251,825 hectares of pine forests in the Sierra Juárez and the Sierra Sur, while a smaller area of 163,784 hectares, mainly in the Sierra Sur, was awarded to the company *Forestal de Oaxaca* (Presidencia de la República 1958a, 1958b). In return for their concession rights, the companies agreed to invest in road building and processing machinery and to pay the salaries of government inspectors. FAPATUX, by far the better capitalized of the two companies, built a pulp processing mill in Tuxtepec in the nearby Valle Nacional. The companies also employed the foresters and forestry technicians. From the point of view of the state, therefore, a key obligation of the logging companies was to introduce scientific forestry and to systematically map and order forests. The concession agreements specified a timetable for the companies to produce a stand map based on aerial photographs, delimit the boundaries of the concession, and finally produce a management plan containing an inventory, cutting plans, and growth calculations in order to regulate "the rational exploitation of the forests"

(Presidencia de la República 1958a). FAPATUX was responsible for introducing order and development into the imagined chaos and under-development of the Sierra, for making forests legible, and for erasing the tangled webs of clientelism among state elites, logging companies, and rural communities.

All of this was, in the end, much too simple and not nearly as easy as the concession agreement described. Although the logging concession theoretically deprived communities of the right to exploit or benefit from their forests, they still had to be placated and paid. On paper, FAPATUX paid the owning community a fee for the volume of timber cut, but most of this was paid into the *Fondo Nacional de Fomento Ejidal* (FONAFE; National Ejidal Development Fund), which held the money in trust. Communities could only make use of this money for approved communal projects, and they had to apply for their own money, a bureaucratic and paternalistic process that they bitterly resented. To add insult to injury, the fees paid by the logging companies were produced in a secretive bureaucratic/administrative process that grossly underestimated the value of the timber. These administratively established timber prices were insufficient to secure community assent, and FAPATUX had to negotiate additional local agreements that provided concrete material benefits such as a school or a road.[6]

The large industrial logging companies like FAPATUX (collectively known as UIEF) were well connected with national political elites and international investors; in the case of FAPATUX, the wife of then president Adolfo Ruíz Cortinez was one of the shareholders, while a Canadian entrepreneur, George Wise, was the main financial backer. The small logging companies that had hitherto controlled these forests had mainly been connected with regional politicians and bitterly resented their dispossession. A contemporary newspaper editorial probably expresses the hostility of these companies, who ironically, painted themselves as the agents of scientific forestry and environmental protection. This denunciation vividly conveys informed knowledge of precisely how political connections could be used to obtain logging permissions and avoid regulations.

> The forest wealth of this state will be exterminated without pity, because unlike the authorizations issued by the Secretariat of Agriculture this type of exploitation is not rational. The concessions derive from the influence of some politician, in such a way that they are authorized to sign contracts with the affected communities, without making inspections or previous studies of the available volume of timber. . . . (Anonymous 1958a)

Widespread industrial logging in the Sierra Juárez was made possible by the completion of the road from Oaxaca to Ixtlán and Tuxtepec in 1958. This road was required by official projects of forest industrialization as specified in the timber concession (Various 1962), but it was also demanded by indigenous communities. Contemporary newspaper accounts describe how the communities of Ixtlán and Guelatao used the symbolism surrounding Benito Juárez's birth in Guelatao to demand state assistance with road building (Anonymous 1953b, 1954). Some communities were so anxious to have the road built that they donated their communal labor crews. Older people in Ixtlán told me that the highway had been actively solicited by community leaders and even accused Guelatao of attempting to route the highway away from Ixtlán. At present, however, most people describe the road as having been imposed on unwilling communities in order to extract timber: In this, they agree with government foresters and ironically even with the governor of Oaxaca. Road building is of course one of the favorite development projects of modern states; thus, the community reinterpretation is a critique of past development that seeks to distance the state by blaming environmental degradation on excessive development. The communities' account of a road built to pillage them of their wealth is of course very largely true (FAPATUX did extract large amounts of timber from the forest). What is significant is that the communities have been successful in imposing their reinterpretation of "development as destruction" on forest service officials who sponsored a video that retold road building as destruction (Grup de Estudios Ambientales 2001). In 2001, even the governor of Oaxaca, José Murat Casab, adopted this explanation when he told a delegation of Sierra community representatives that he would not repair the road because it was built to take away the communities' wealth and not to help develop them.

Forestry Meets Resistance: Negotiations over Logging

FAPATUX's efforts to introduce industrial logging into the Sierra Juárez met with resistance. Some communities had already experienced officially approved selection logging systems, which the forest service had been promoting since the 1930s and which had less visible impact on the forest. When FAPATUX began logging in 1958, it attempted to introduce a clear-cut logging system but was defeated by local opposition, partially inspired by a desire to keep logging jobs in the hands of local people, but also partially by a critique of the clear-cut system. Community leaders used the theories about selective logging sponsored by the forest service in order to oppose the technologies and clear-cut logging practices of the paper company.

The clear-cut logging system was based on Canadian cable logging technology: It required a good network of roads that would allow truck-mounted logging towers to stretch winch cables outward into the forest. This was a high-technology solution, which required highly skilled workers who knew how to manage and maintain towers and cables. Initially this skilled work was carried out by loggers from the states of Michoacán and Jalisco, where industrial logging had a long history. In 1958, FAPATUX built a logging camp for one hundred of these out-of-state workers in Llano de las Flores, in lands belonging to the community of Atepec, north of Ixtlán. Although the clear-cut system was potentially well suited to the ecological requirements of the pine species in this area, the apparent devastation it caused was unacceptable to the community of Atepec because cuts visibly "killed the young trees" (Gutierrez 1986).[7]

In Atepec in 1958, clear cuts were unacceptable to the community because they did not accord with the selective logging guidelines that the forest service had been promoting since the 1930s, because they employed outsiders rather than community members, and because they appeared environmentally devastating. Community leaders petitioned the federal government to ensure that all logging jobs be given to community members (Escarpita Herrera 1980), and this policy was subsequently applied in the rest of the Sierra Juárez. The employment of community members as loggers effectively caused the abandonment of the tower logging system, which could have pulled large logs intact to the roadside. In contrast, community members cut logs to manageable 1.3-meter lengths and brought them to the road by either rolling them downhill or debarking, splitting, and manhandling sections uphill. On the steep slopes of the Sierra Juárez, this was a dangerous and laborious process even after the introduction of cable winches in the 1960s. A direct result of forest communities' success in imposing a selection logging system was the much greater number of people involved in logging and timber extraction who were exposed to FAPATUX's logging plans, timber marking practices, and scientific theories. The selection logging system had the paradoxical effect of reducing the value of the timber to the logging company; large-diameter trees of great potential value as saw-timber were converted into short lengths that could only be used for low-value paper pulp.[8] On the other hand, the employment of *comuneros* as loggers provided a greater direct benefit to the communities themselves. A political and economic conflict between the company and the communities had been framed as a conflict over the technical, a disagreement about the rationality and environmental impacts of a logging technology. Although the communities had not openly won the political conflict as to who owned the right to exploit the forest, they had won a significant victory by transforming a political conflict over forest resources into a technical argument about logging technologies and work practices.

Atepec's success in imposing a selection logging system on FAPATUX had ecological and economic impacts; on the one hand, the community had succeeded in retaining more of the economic benefits of logging by insisting that all logging had to be carried out by community members. On the other hand, the selection logging system favored an increase in oak species at the cost of commercially attractive pine, while it removed the most valuable trees, so that the volume and quality of standing timber in the forests was steadily reduced over the duration of the concession (Snook 1986; Chapela and Lara 1993). Another feature of this episode was Atepec's determination to claim control of forests by insisting that community members should do as much of the work as possible. This was to be a consistent goal of forest communities. Beginning with logging in 1958, community members took control of timber transport and the logging trucks during the 1960s, and during the 1970s and 1980s, a number of *comuneros* in Ixtlán and other leading communities were trained as forestry technicians or foresters, sometimes with community sponsorship. As a result, when the FAPATUX logging concession was canceled in 1985, the most important forest communities of the Sierra had a pool of trained loggers, truck drivers, and forestry technicians.

The logging concession has had an enduring impact on the structure of forest stands, on species distributions, and on work in the forests. Job descriptions and logging practices established during the concession have persisted largely unchanged to the present day, road networks established under FAPATUX have allowed logging that has gradually altered the age distributions and sizes of trees, and the species composition of forests. The most important division in the labor force, however, was between FAPATUX employees[9] and community members. Company employees were better paid and received health care and benefits, whereas community members were paid by the volume of timber they cut.[10] Relations between FAPATUX and community loggers were marked by suspicion and mutual cheating. Loggers would "accidentally" cut extra trees or seek to have timber double counted, whereas FAPATUX logging technicians are widely believed to have undermeasured timber so as to line their pockets.

Economic and Ecological Impact of Logging, 1958–1982

Pine forests were also actors in this drama: It was after all pine trees that were being cut, dragged to roadsides, and shipped away to make paper. Between 1958 and 1982, almost all of the accessible pine and pine-oak forests of the Sierra Juárez were logged at least once. Although this did reduce the economic value of the forest through the removal of the largest and most valuable trees, pine forests were well adapted to this limited disturbance, and little deforestation would have occurred as a result. The selection logging system operated by FAPATUX overwhelmingly concentrated on large diameter pine trees, causing an expansion of oak presence in logged stands and leaving behind smaller and often damaged pine trees. A powerful indirect effect of logging was the increasing abandonment of agriculture in many Sierra communities, resulting in the wholesale colonization of abandoned fields by young pine trees. Pine regeneration in established forest, on the other hand, was diminished by successful community fire suppression and by the selective logging system; with the increasing rarity of fires, pines could only regenerate on heavily logged areas or after rare large wildfires, as at Cerro Pelón in 1973 (Zorrilla Sangerman 1973). In summary, forest area advanced even as many parts of the forest suffered the impact of selective logging; in some areas pine trees advanced to occupy abandoned fields, whereas in other areas oaks expanded to take advantage of spaces offered by the removal of pines. The impact of the logging concession was certainly not to eliminate forests: On the contrary, unruly pines and oaks expanded to occupy the spaces offered by official logging practices, unofficial practices of evading logging rules, and unforeseen effects of suppressing forest fires.

This was a complex panorama of environmental change, but people recounted this period as one of unmitigated environmental degradation. People in Ixtlán spoke of the environmental degradation caused by the logging company in terms of timber, describing the extremely large and ancient trees that they believed had been present before the arrival of FAPATUX. Pedro Vidal García Pérez, a *comunero*/forester from Ixtlán, described this environmental degradation in a book that used old forest service documents to calculate that average diameters had declined from

1.3 meters to 90 centimeters by the 1970s (García Pérez 2000). The tradition of scholar/advocates of the community of Ixtlán, is alive in such claims. My personal assessment is that these claims may be exaggerated by some quirk of selective logging,[11] but in any case it is likely that even very large trees were probably relatively young and rapidly growing *Pinus patula* from the moister Atlantic side of the watershed. It is difficult to assess the ultimate impact of FAPATUX era logging upon the forests: initial inventory volumes were sketchy and optimistic and probably inflated in order to attract investment (e.g., FAPATUX 1956). Present day estimates of the impact of the concession have tended to take these over optimistic estimates at face value, yielding inflated claims of economic damage (Chapela and Lara 1993). Nevertheless, it is clear that logging decreased the value of standing timber and left damaged residual stands. From a purely economic point of view, FAPATUX was a consistent failure. The company had trouble making a profit, and the conversion of high-value timber to low-value newsprint was an extraordinary destruction of possible economic value, whether to the forest communities or to the logging company.[12] By 1976, annual industrial timber cutting in the state of Oaxaca had increased from the relatively modest levels of about 100,000 cubic meters during the 1940s and 1950s to 524,000 cubic meters, of which 178,000 cubic meters was pine timber cut for pulp in the Sierra Juárez (Tamayo 1976). Industrial logging in this relatively small area had far surpassed traditional fuelwood cutting and charcoal burning for local consumption, which was by then perhaps 50,000 cubic meters per year.[13] This was a dramatic shift from the situation as recently as the 1950s, when by far the dominant use of forests was for fuelwood. This was the beginning then of the displacement of fire from landscapes to internal combustion engines, making it possible for pine timber to travel farther than ever before, even as oak cutting for firewood began to be displaced by fossil fuel stoves.

The Struggle Against FAPATUX and the Rise of Community Forestry

Between 1958 and 1982, the forest communities of the Sierra Juárez became increasingly disenchanted with FAPATUX. Relations between the communities and the company were perennially troubled; the communities complained about poor pay and labor conditions and about the difficulty in getting their money from the FONAFE. In 1967, this discontent boiled over into a timber suppliers' strike, initially led by the community of Macuiltianguis but subsequently joined by Comaltepec,

Atepec, Ixtlán, Aloapan, Ixtepeji, Luvina, and Teococuilco. FAPATUX's initial response was to seek timber from communities that had not yet participated in logging and to attempt to divide the striking communities by a combination of threats and improved offers to individual communities and community leaders (Martínez Luna 1977). The core group of communities continued to strike until 1972, when the paper mill in Tuxtepec was closed for forty days (Bray 1991). Ultimately, FAPATUX agreed to increase rates of pay and improve working conditions in the forest. For the core striking communities such as Macuiltianguis, the success of the strike came at great social cost; the long strike had undermined the economy and exacerbated pressure to emigrate. The strike laid the basis for future alliances between forest communities against the company, but it was also a divisive event that was the source of bitter feelings between and within communities. People from communities like Macuiltianguis, which maintained the strike for the longest, remain distrustful of communities like Ixtlán, which returned to work sooner.

Under a new director, Jorge L. Tamayo (1974–1979), FAPATUX attempted to counter the power of striking communities by organizing a massive increase in timber production from hitherto untouched areas of forest (Tamayo and Beltran 1977; Tamayo 1976) and by building community support by creating two "community" saw mills, in Ixtlán de Juárez and Concepción Papalos.[14] The Ixtlán saw mill provided jobs and training to community members, but communities that supplied timber to the mill felt that most of the profits went to Ixtlán and suspected, correctly, that timber prices were being manipulated in favor of the logging company. In addition to these paternalistic policies, FAPATUX may have colluded in coercive tactics as when activists opposed to logging in Comaltepec, Quiotepec, and Yolox were arrested by the army in 1979 (Szekely and Madrid 1990). Whether FAPATUX was directly involved in these arrests, the association between counterinsurgency and logging is a persisting, if subdued element in accounts of this period.

The logging concession was due to expire in 1983, but as the concession neared its end, FAPATUX began to apply pressure on the government to renew or extend the concession indefinitely. The forest communities were of course well aware of the paper company's efforts and organized to press for complete cancellation of the concessions, arguing that they could manage their own forests, and that the logging company had been environmentally destructive and corrupt. A highly significant role in informing and then supporting the communities in their fight against the concessions was played by outside activists and a burgeoning NGO movement. One such group surrounded Jaime Luna, a leftist intellectual who had moved to Guelatao in the early 1970s and founded the NGO *Comunalidad*. In 1979, Luna and his group helped found ODRENASIJ, an alliance of forest communities opposed to the logging companies.[15] Luna and his collaborators in the ODRENASIJ group were highly influential in producing pamphlets opposing FAPATUX (Anonymous 1983b); they helped monitor the legal proceedings surrounding the proposed concession renewal, and they participated in the mobilization of community protests. In the Sierra Sur, similar protests opposed a new concession to the *Compañía Forestal de Oaxaca*. Although the concessions were extended by the outgoing president, José López Portillo, in November of 1982 (Ortega Pizarro and Correa 1983), massive popular protest and unfavorable national newspaper attention resulted first in the suspension and then in the cancellation of the concessions by the new president, Miguel de la Madrid (Leon 1983; Muro 1983).

This was a heroic period for the communities and activists involved: At present, almost everyone in Oaxaca who was involved in forestry claims to have been on the side of the communities. This is, however, a contentious legacy as people disagree bitterly over the relative contributions of different activists, NGOs, communities, and government officials. The symbolic value of this narrative of illegitimate depredation, protest, and ultimate community triumph is of great value in current environmental politics, a moral high ground that many are anxious to claim. Ironically, it is a narrative that is even deployed by communities such as Ixtlán, which was reluctant to oppose FAPATUX and only joined the anti-concession movement once its success seemed assured. The legend of united communities opposing state and logging companies has similarly been successfully diffused by several of the articulate and intelligent voices in the anti-concession fight (Martinez Luna 1977; Chapela and Lara 1995). It is this version that commands national and international media attention and often brings international researchers such as myself.

Inevitably, this political narrative of community unity and steadfast resistance has obscured the alliances among FAPATUX, the forest service, and forest communities. A complex process of negotiation, accommodation, and resistance is but one episode in the longstanding tradition of alliance and opposition between state and communities, not only in Oaxaca, but elsewhere in Mexico. Celebratory narratives of unity or condemnatory narratives of betrayal misrecognize the political motivations of the Sierra Juárez communities that decided to join or ignore the anti-logging movement, and they conceal the longstanding tensions between and within communities. For communities such as Ixtlán, which had already negotiated a successful accommodation with FAPATUX, it made little sense to abandon their alliance with the logging company. For Ixtlán's weaker and less well-connected neighbors, always suspicious of its power and authority, community forestry NGOs such as Jaime Luna's group provided a new set of possible alliances through which outside institutions could be approached. These marginalized communities were correspondingly more active in ODRENASIJ, which collapsed soon after the cancellation of the logging concession in 1983. For most communities, ODRENASIJ had fulfilled its purpose, and they could now turn to their next objective: gaining direct control over forest management and the profits it could provide for the community. The primary focus for *serranos* has remained the community. Intercommunity umbrella organizations have had very limited success, except when, as in the movement against the logging company, they respond to outside forces that community leaders believe they can more successfully fight as a coalition than as individual communities. Such alliances are usually temporary because smaller communities are suspicious of historically powerful communities and because a classic mode of cooption by the Mexican state has been to form opaque umbrella organizations of communities, businesses, or peasant organizations. Umbrella organizations can be oppressive, corrupt, or even dangerous.

Indigenous Identity and the Struggle for Control of the Forests

The indigenous opponents of the logging company did not call on indigeneity, nor did they claim to be environmentalists. Their campaign was for dignified and well-paid work, not a claim to be indigenous protectors of the environment. In contrast, for the national media, the anti-FAPATUX struggle was described in terms of the specifically indigenous identity of the communities involved (Anonymous 1983c, 1983d; Leon 1983). Subsequent international attention has also focused on the indigenous character of the movement (Bray 1991), while much of the coverage highlights the specifically environmentalist goals of these protests. From the point of view of *serranos* themselves, indigenous identity or environmentalism were much less important than a specifically communal identity and a demand for just compensation. Contemporary ODRENASIJ pamphlets make no mention of indigenousness or environmental degradation

(Anonymous 1983a, 1983b), and many ODRENASIJ supporters framed the conflict around pay and labor conditions and described the communities as being made of peasants rather than indigenous people (Chapela and Lara 1995). Here, the categories of Marxist political economy came to describe indigenous people who had yet to think of themselves as indigenous or as environmentalists in a political register. People from Ixtlán never retold me their struggle for control of their forests using the language of indigeneity, preferring always to describe the conflict as one between the *community* and the logging company. Some neighboring communities do frame political identity in indigenous terms,[16] but it is not at present a political category that solidifies broad alliances between communities. It is of course possible that indigenous identity may yet emerge in the context of alliances of communities in the Sierra Juárez. However, given that these alliances have tended to be short lived when compared with the solidity of community institutions, it seems unlikely that a regional indigenous identity will be an important political category for the immediate future. The emergence of indigenousness as an important discourse in national politics since the 1994 Zapatista rebellion could provide the Sierra Juárez communities with a set of outside institutions with whom they may wish to form alliances. In such a situation, indigenous identity could become more salient.

The reluctance of communities to form intercommunity organizations is demonstrated by the history of forest management from 1983 onward. After a brief hiatus in logging immediately after the cancellation of the FAPATUX concession,[17] forest communities were in theory allowed to sell their timber to the highest bidder. The forest service somewhat replaced the logging company regime by requiring forest communities to pay forestry technicians from various umbrella organizations, effectively restricting communities' ability to employ their own foresters and sell their own timber. In 1992, a new forest law set the communities free from these restrictions, allowing them to hire their own foresters and order their own management plans. The largest communities swiftly pulled out of umbrella organizations whether state sponsored or NGO supported. As in the case of ODRENASIJ, the communities were reluctant to be involved in any umbrella organization once the political or legal reasons to do so were removed. At present, forest communities can hire their own foresters: Relationships between foresters and communities are typically individualistic and market based, but personal relationships forged during the 1980s continue to affect which forester works with which community.

Land Use Change in Ixtlán, 1945–2001: The Decline of Agriculture

By the late 1930s, farmers around Ixtlán depended on the mine of La Natividad as the principal market for their produce, and the decline in mining after 1945 contributed to a rapid decline in agriculture. Many of these farmers abandoned producing food for local markets in order to work as migrant laborers in the United States or as wage laborers in the logging business, which arrived in Ixtlán in 1948 (Epifanio Pérez, interview notes, July 25, 2002). This decline in agriculture was further accelerated by the construction of the Oaxaca–Ixtlán highway in 1956, extending to Tuxtepec in 1962. The highway delivered a death blow to the traditional mule caravans. Although the rise of coffee cultivation in the warmer tropical areas around Villa Alta increased the purchasing power of communities in that area, their money did not flow into the regional agricultural economy. Cash earned from coffee cultivation was used to buy food and manufactured goods imported from Oaxaca, Veracruz, and other parts of Mexico. As a result, the area of agricultural land in the district of Ixtlán declined from 22,973 hectares in 1950 to 4,789 hectares in 1980 (Abardia Morelos and Solano Solano 1995). Although this government census data should be treated with some caution, aerial photographs of Ixtlán confirm a dramatic increase in forest between 1967 and 1995 (see figures 5.1 and 5.2). The 1967 aerial photographs show fields (white areas) that are forested (dark areas) in 1995. The difference between the two pictures is all the more dramatic when we consider that much of the decline in agriculture had *already* occurred by 1967: The area in cultivation in 1945 was perhaps four times larger still than in the 1967 photograph.

The remaining farmers in Ixtlán have increasingly concentrated on productive irrigated and permanently cultivable land near town, eliminating the need for them to use fire as an agricultural tool in rainfed *milpa* fields dispersed through the forests. Settlement patterns have also changed as people have moved to town to work as loggers or saw mill workers, although a few loggers cultivate fields near their homes in town on weekends. In the Sierra Juárez more broadly, the period of greatest agricultural abandonment between 1950 and 1980 was marked by heavy out-migration, causing overall population levels to remain broadly stable in spite of relatively high birth rates.[18] Local agricultural production for subsistence and trade was increasingly replaced by food imported from other parts of Mexico, creating increased dependence on wage earning, cash crop production (in the coffee growing zone around

Figure 5.1
Aerial view of Ixtlán de Juárez in 1967. *Source:* Photograph courtesy of Servicio Geologico Mexicano.

Villa Alta), or, more recently, remittances from relatives in the United States.

Fire Becomes Wild as People Become Urban: The Decline of Agriculture and the Rise of Forestry

As we have seen, FAPATUX recycled official ideology of fire suppression in its self-serving justifications, but it is unlikely that it ever did much actual fire fighting. Caring about fire, fighting fires, and talking about past fires have come to be ways that people in Ixtlán understand themselves as protectors of the forest, marking a radical transformation from the kind of technological knowledge of controlled burning that had formerly been widespread. Although the company built lookout towers to give warning of fires, people in Ixtlán were adamant that it was the

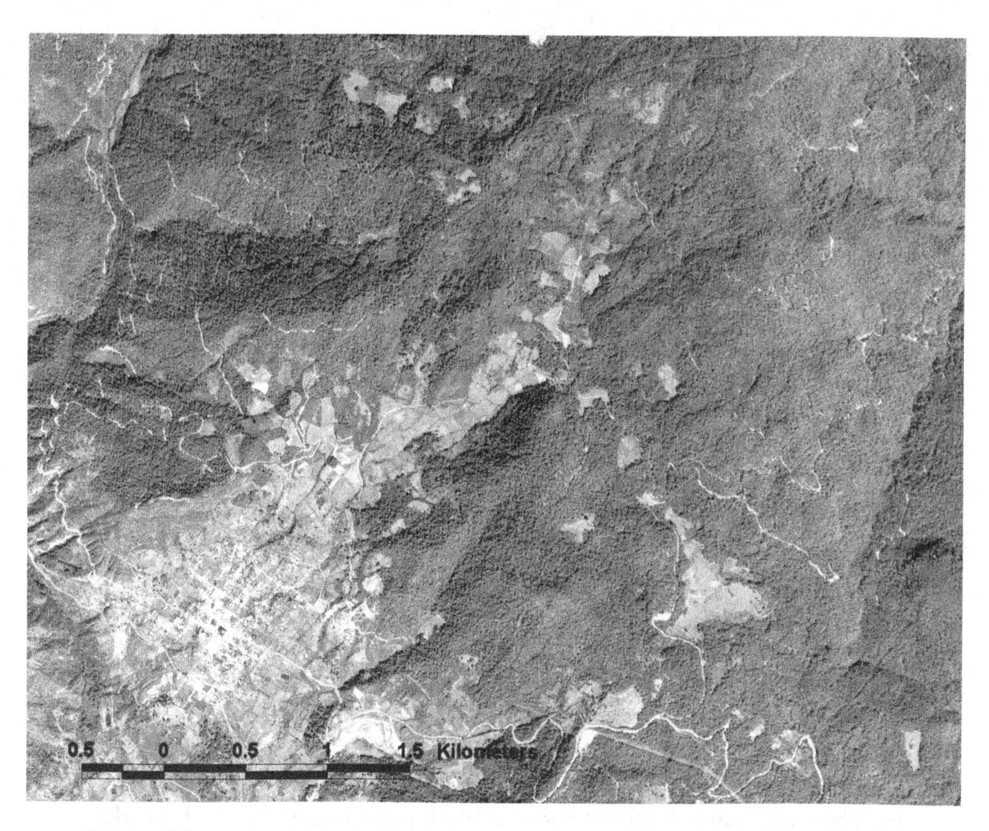

Figure 5.2
Aerial view of Ixtlán de Juárez in 1995. *Source:* Data from INEGI 1995. Figure by author.

community that actually coordinated and carried out fire fighting. Although FAPATUX did seek to regulate *milpa* agriculture with its associated use of fire, people in Ixtlán told me that this was because company foresters feared that people might use cultivation as a pretext to cut and sell pine timber (Jose Pozo, author interview notes, August 3, 2000). Fire fighting was carried out by the community as a continuation of traditional fire control measures that were integral to swidden agriculture and pastoralism, but thanks to logging roads built by FAPATUX, community fire fighters could quickly reach remote areas that could not be reached in the past. Fighting forest fires was therefore a continuation of traditional fire control practices in the new context of industrial forestry, spreading out from the roads and logging areas established by the timber company. More significantly, the company provided a new set of theories according to which fire was destructive because it threatened commercially valuable pine timber.

Fire was an integral part of traditional *milpa* agriculture, as described to me by the older men of Ixtlán. They talked of fire knowledgeably; they knew fire as a tool and an unruly ally. They agreed that agricultural fires could escape into neighboring fields or forests through carelessness or bad luck, but that fires could usually be controlled through the use of firebreaks and by careful burning. Skill was something others could assess, and clumsiness could be punished by the community:

> Even when I was young, if people burned without giving notice, if they didn't open up firebreaks and if the fire got away they would be punished with [the community] jail. (Zenaido Pérez, author interview notes, July 24, 2002)

In the context of *milpa*, fire was domesticated; it was kept within cultivation, and a system of norms regulated its use. The decline of *milpa* agriculture implied a corresponding decline in the domesticated uses of fire, while the rise of industrial forestry expanded the arena where fire was *outside* cultivation, where it was wild and destructive.[19] As more and more people made their living cutting timber in the forest, fire came to be seen as increasingly destructive and wild, and the more benign traditional uses became submerged: One kind of knowing displaced another. At present, fire has an important role in defining community membership; *comuneros* are responsible for fighting fires and have the right to receive benefits from the logging business. People from Ixtlán see their willingness to control agricultural fires as distinguishing them from people in neighboring communities, who they describe as either malicious and ignorant (because they do not set fire breaks around *milpa* and allow fires to escape) or incompetent (because they do not know their own forests well enough to fight fires successfully). For the purposes of distinguishing themselves from their neighbors then, people from Ixtlán have adopted the state ideology according to which agricultural fires are spread by ignorance and carelessness.

This shift in the meanings of fire has caused a shift in people's identities; they have moved from defining themselves as people who know how to *use* fire to people who know how to *fight* fire. Crucially, however, the theories of fire promoted by the logging company have been applied to make new knowledge of fire and new identities for themselves as the opponents of fire. Far from being a theory that people internalized in order to police their own actions, fire suppression was a theoretically informed practice that produced solidarities and political possibilities in relation to the state. Older *comuneros* still remembered their fire using identity and the theories and practices of *milpa* agriculture, but for younger people who grew up during the forestry concession and after the abandonment of agriculture, it is their identity as fire fighters, forest workers, and loggers that is salient. They particularly defined their willingness to fight fire against the forest service (which does nothing) and neighboring communities that allow fires to escape cultivation or do not fight fire properly. For the generation of *comuneros* presently in power, fire fighting and the willingness to fight fires are both a responsibility and a mark of pride and difference; because only *comuneros* can work as loggers, willingness to fight fires also defines who can profit from forestry activities.

The decline of *milpa* agriculture does not seem to have been caused directly by community or paper company sanctions; even José Pozo, an old *comunero* who was otherwise a scathing critic of the community forest business, never accused it of preventing farming. Rather, older men told me that people abandoned agriculture because they wanted to earn money in forestry (Arce, author interview notes, September 14, 2000). With the rise of forestry, the *Comisariado de Bienes Comunales* has come to be responsible for protecting forests from agriculture. Access to land for agriculture had formerly been relatively unregulated, but as the *Comisariado* consolidated its control of forests, it began to require people to ask permission before clearing forest for *milpa*. The central aim was to protect pine trees that would one day have commercial value.

> Now it has changed a lot, the situation is very different from the past, in order to ask for new land you have to ask for permission, but [the *Comisariado*] only give it in certain situations, they send to see. If there is a lot of pine they don't give permission, but if there is tropical vegetation or oak they do. (José Pozo, author interview notes, July 25, 2002)

Comisariado control of agriculture has been implemented primarily through direct control of people's right to cut trees even on reforested fields; although ownership of land that has once been farmed is recognized, people may not necessarily be allowed to cut trees even on this land. As the *comunero* Zenaido Pérez told me:

> Now the fields have reforested, and no one can fell trees. Before anyone could cut where they wanted, now they have to ask for permission from the authority. . . . Now it is forbidden to cut [*rozar*], those who want to cultivate their land don't allow it to reforest. (interview notes, July 24, 2002)

People remember where their fields used to be and feel a remaining sense of ownership even if they no longer farm there. The landscape still bears some legacies of past days of burning, planting, and harvesting in the thriving young stands of pine trees that occupy former fields.[20] Zenaido, who worked for many years as a logger, had adopted some of the techniques and language of forestry into a new form of caring for young trees, talking of how he thinned young pine trees on his former fields. This was a hybrid practice indeed, an old swidden farmer returning to his swidden to kill some trees and favor others, to welcome the vigor of young trees and to talk about these activities using the language of forestry. Here, as with fighting fire, scientific theories were not imposed on Zenaido; rather, they were a resource for encountering trees and for knowing himself as someone who cared for trees. Unruly trees became a resource for making identities, making a living, and making political claims and alliances.

Community Forestry: Resistance, Hybridization, or Appropriation?

It would be hard to argue that a state- and company-sponsored ideology has been imposed on rural people or that official environmental theories have become a kind of regime of discursive self-policing. There is here, on the contrary, a creative reworking of official discourse and history, a making of knowledge about forests and histories of environmental change. People in Ixtlán have taken elements of the environmental narrative produced by FAPATUX and the *Comisión del Papaloapan* in the 1950s and reworked them into an environmental history that justifies community ownership of the forest. Community leaders told me that FAPATUX and other logging companies had pillaged a forest that was largely untouched, taking the largest trees and leaving only the smaller ones behind. Zenaido Pérez recounted these events to me in 2000:

> No one had exploited the forest, it was almost virgin, although there were a few people who sold planks, but very few. He [the logger Manuel García] took the largest trees, the best of what there was. I remember one day, working with a two man saw we cut a tree 6 feet across, it took us half a day to cut it, it was a father tree . . . now we really suffer the consequences of *irrational exploitation*[21] [i.e. by the logging companies]. We need to fight forest fires to protect the forest. Fire can destroy the forest, but we here have the advantage that the forest reforests by itself, we have seen that. But the forest gives to us, and we have to think, what are we giving back to the forest? (interview notes, July 25, 2000)

Zenaido inverted the 1956 FAPATUX description of the forest as degraded, claiming that the forest was "virgin" in 1948, that it was the logging companies and not the communities that had degraded the forest, and that the communities had the right and the responsibility to restore this degradation. His and other *comuneros*' denunciation of present-day degradation is an account of how they see forests (in terms of timber volume), of themselves as careful knowers and workers in the forest, and a political claim to restore and control the forest. Zenaido's account shared the 1956 FAPATUX definition of degradation in terms of timber volume and fire and reworked it into the present-day discourse of degradation used by environmentalists, forest service officials, and community leaders.

Imposing Degradation Histories on the State

The shared public fact of environmental degradation has been imposed on the forest service by Ixtlán and its allies. This is a coproduced public knowledge at the interface between state and community, a knowledge that defines who knows forests and what kinds of institutions rest on this knowledge. The forest service and the forest communities of the Sierra Juárez are in close agreement about the history of degradation of the forest, the terms in which degradation is to be defined, and the practices needed to regenerate the forest. Both the forest communities and the forest service view fighting forest fires as a duty, and both parties publicly agree that *milpa* agriculture, which depends on long forest fallows and the use of fire, is reprehensible. The forest service officially supports the view that FAPATUX pillaged the forest of its largest trees and has even sponsored books and videos that attribute deforestation to the logging company (Grupo de Estudios Ambientales 2001; García Pérez 2000). It is worth pausing to consider how dramatic an inversion this is: just compare the new authors of degradation with the authors and causes of degradation identified by FAPATUX and the *Comision del Papaloapan* in the 1950s, and you realize what a profound change has taken place. This new official history has been imposed on a forest service that has to collaborate with forest communities; most officials feel no conflict with this new official ideology because few of them are personally associated with past policies. The official history of environmental degradation, which blames FAPATUX, acts as a "boundary object" (Van der Sluijs, Shackley, and Wynne 1998; Star and Griesemer 1989) that links the social worlds of forestry officials and members of forest communities, anchoring a language of timber volume, forest management, and fire fighting while allowing forest communities considerable autonomy from state supervision. Crucially, the political space of forest management has created a shared political identity for community members; this identity enables forms of collective action that limit the power of forestry officials through such actions as public protests and road blockades.

Traditional Agroecological Knowledge as a Critique of Modern Forestry

Although *milpa* cultivation has largely been displaced by the ideology and production system of industrial forestry, it persists as a coherent albeit marginalized system of theories and practices from which criticisms of forest management can be mounted. It is striking that such criticisms ally themselves with the power and generative possibilities of fire to farm fields and transform forests. The most sustained criticism I heard of present-day forest management practices in Ixtlán came from an 83-year-old *comunero*, Jose Pozo, who used his knowledge of milpa agriculture to question the value of community reforestation projects:

> My nephew studied to be a forestry technician, and I asked him, why did the [community foresters] study . . . they waste money planting trees in the forest. I said to him, "What are you doing there wasting money planting trees, you should spend the money on our needs. You have education but you don't know that those forests don't need to be reforested [i.e. planted by people], these forests reforest themselves. You should go to the tropical zone with your companions, where there is no pine, only a few large ones. You need to cut there, make a *milpa*, harvest two or three years of corn, and then when the soil is tired the pines will be born there because the wind will carry the seed there. When the land is no good for maize any more you can leave it and the pine will be born there." (interview notes, August 3, 2000)

For Pozo, the forest was more powerful than for the community foresters, who he accused of ignoring the regenerative powers of forests in the face of fire. From the perspective of *milpa*, agriculture fire is not necessarily destructive: As Pozo pointed out, old burns would rapidly be covered with dense pine regeneration. This point of view was definitively marginal within Ixtlán. The director of the community forestry business, Leopoldo Santiago, had experimented with prescribed burns to stimulate pine regeneration, but in several instances fires nearly escaped, exposing him to bitter community criticism. As a result, the preferred means of regeneration remained manual replanting with nursery grown tree seedlings, which Pozo considered a waste of time. Tellingly, the only context where fire is an acceptable tool is to burn trees affected by the *Dendroctonus* pine beetle, which can destroy thousands of hectares of valuable timber and cause large fires in the dead forest.

In Ixtlán, the knowledge system of traditional agriculture, where fire was potentially benign, has been submerged and overwritten by the knowledge system of modern forestry, which was and is inimical to *milpa* cultivation. There is no necessary opposition between these two knowledge systems; the literature in the sociology of science is replete with examples of creative hybridization between different knowledge systems, and under other circumstances, some kind of accommodation between these two systems of knowledge could conceivably have occurred. As Akhil Gupta points out, traditional farmers and indigenous people have a long history of creatively appropriating knowledge of all kinds,[22] and indigenous ecology is itself a hybrid (Gupta 1998:290). In fact, the metaphor of hybridization is in some ways misleading, suggesting a natural process of growth and the mixing of inherently separate varieties. A more accurate metaphor would draw attention to the intentional practices of real people who mobilize knowledge and produce or make use of concepts of difference such as indigeneity or modernity. Clearly, knowledge does not hybridize by itself, rather it is mobilized by specific people who learn it and try it out for specific purposes and who have considerable freedom about what kinds of knowledge they can adopt. These mobilizations of knowledge do, however, take place in the context of uneven power relations, which can push knowledge and identities on people and to some degree impose it against their will. In the Sierra Juárez, the forest service and FAPATUX sought to impose industrial forestry and to suppress *milpa* cultivation. As I have shown, FAPATUX was ultimately unable to keep control of the forests, but it was more successful in introducing its knowledge about forests to indigenous people. In the context of Mexican forestry in the twentieth century, forest science was fervently opposed to swidden cultivation, and this opposition has been entrenched in the institutions of community forest management. Community and state institutions have stabilized knowledge of forests as a repository of timber, marginalized knowledge of forest fires, and suppressed the dramatic fire history of the forests of the Sierra Juárez. Just as the institutions of community government are a product of past encounters between the colonial state and the indigenous people of the Sierra, so too the new forest management organizations are the product of the encounter between the modern Mexican forest service and the forest communities.

Discourse, Power, and Memory

The official discourse of fire and the separation of agriculture and forests has become part of the worldview of the people of Ixtlán. Community members have come to take the separation of agriculture from forestry for granted; they know that abandoned fields have regenerated abundantly with pine, but the dominant discourse of fire as degrading gives this knowledge no interpretive framework, and there is no public space where it might be discussed and acted on. The separation between agriculture and forest was first made by the forest service in the 1930s. It is now part of the organization of community life because work in the forests and the fields is separate in both theory and practice; no one at present works long fallow *milpa*, so there is no land moving from forest to agriculture and back. Individuals may know that old fields are now pine forest, but this does not fit the discourses of degradation and forest management used to write management plans and claim control of the forest from the state.

Memory of past fires has also been suppressed by the creation of new knowledge of fire through the practice of fighting fires and the production of the political identity of fire fighting. In this sense, state fire control discourse allowed the creation of a new political and experimental space where *comuneros* have encountered nature, where they have made knowledge of what forests are and who they are. This new political identity has allowed people to make creative use of official fire control discourse in order to produce community political power. Fire fighting takes place in the context of protecting the standing timber in community forests—only those who are willing to fight fires can become full *comuneros* with the right to participate in the community forestry business. Their difficult, dirty, and dangerous encounter with fire has led community members to see fires as the opponent of both their forests and their livelihoods. They know about fire in a direct and active way: how fires in the dry pine oak forest tend to burn through the under story without becoming crown fires, how fires in the moister *P. patula*-dominated forest can become intense stand-destroying fires, and how to cut a firebreak and set a backfire to deprive an advancing fire of fuel. This detailed knowledge defines fire only as a danger and a threat; it is knowledge that submerges memories of more benign uses of fire by farmers and pastoralists in the past. Community members' present-day knowledge of fire is conditioned by the social world of organized logging and timber extraction and is framed with the theories of industrial forestry: Identities, practices, and theories combine to produce knowledge of forests and forest fires. The general degradation discourse that the community has adopted, modified, and reflected back on the state provides part of the interpretive frame for knowing fire and forests, but it is the community political organization and practices of commercial logging and fire fighting that are critical in defining how *comuneros* come to know fire and forests. Knowledge and political order are coproduced and stabilized against each other (Jasanoff 2004) even as popular memory of fires is suppressed.

Over the last fifty years, then, successive state interventions into the Sierra Juárez have produced popular knowledge of ways to build conceptual and political links between the forests of the Sierra Juárez and the wider world. As we shall see, the theories about the relationships among forests, floods, and water supplies first propounded by the Papaloapan Commission have come to provide a symbolic resource for community members who wish to bypass community and state forestry institutions and build alliances with national and international actors in conservation. The communities of the Sierra Juárez are cited as successful examples of forest management (World Bank 1997; Bray et al. 2003; Szekely and Madrid 1990); they are also examples of the triumph of fire suppression over the agricultural use of fire. As swidden agriculture begins to disappear in other parts of the world and as fuelwood burning gives way to gas stoves in other places, collective understandings of landscapes will underpin new institutions and produce new invisibilities—new blind spots that are hard to see. In the Sierra Juárez, the price of a good working relationship with the forest service is the loss of local knowledge of fire and forest history. Popular memory of fires and an increase in forest area have been suppressed during the struggle to secure community control of forests. Like all forms of knowledge, narratives of environmental degradation sustain some memories and silence others, support some knowledges and suppress others. Categories that separate the landscape into field and forest make invisible the lively movement back and forth between forest and field: It is a good time to be a pine seedling in the Sierra even if no one notices you.

6

The Mexican Forest Service: Knowledge, Ignorance, and Power

In November 2000, I attended a convention on community forestry in a hotel on the outskirts of the city of Oaxaca. For three days, government officials, scientists, and the occasional NGO representative occupied an elevated stage and presented their views on the state of Oaxaca's forests before an audience of indigenous community representatives. Officials and NGO representatives were easy to spot; they got to sit on stage, they participated confidently in debates, and they wore clearly "modern" clothes—polished street shoes, button down shirts, or occasionally the plaid shirts preferred by many urban Mexicans who work in the countryside. Although indigenousness was seldom discussed, most of the people on stage were clearly paler skinned than their audience. All of the community representatives who I talked to spoke Spanish, but many had the characteristic clipped accent often heard in indigenous communities where Spanish was only recently, or not yet, the language of daily life. Some community members were also clearly office workers or professionals and wore the appropriate "modern" clothes, but many wore baseball caps or the plastic-coated straw hats popular among rural Mexicans.

One of the stated goals of the convention was to elicit the views of community members about forest management and forestry regulation in a state where most forests were owned by indigenous communities. However, it became increasingly clear that a much more important objective was to build political support for the forest service, SEMARNAP,[1] and for a World Bank-funded forestry project, PROCYMAF, which sought to increase industrial forestry in indigenous communities. In an inaugural address, the director of SEMARNAP for the state of Oaxaca announced that a principal objective of the convention was: "to let society know that communities protect their forests, that they generate jobs from the forests, and generate environmental services" (fieldnotes,

November 8, 2000). He went on to describe the "advances which society should know," reciting figures about timber production, areas of forest under management, and the numbers of community members involved in fire fighting. Critically, he declared, "Only 2% of burned areas are in communities with forests under management"; violations and environmental degradation took place mainly in unlogged forests. Project leaders and forestry officials were trying to bolster support for logging by producing a representation of community forestry that would visibly demonstrate the political power of the forestry sector to a hostile governor and to his environmentalist allies.[1] Officials wished to enlist forest communities and to stage manage a representation of successful forestry development that could link the pine forests of Oaxaca, the legal boundaries of forest communities, the bodies of the community representatives in the room, and global scale World Bank development agendas. The convention therefore provided a theatrical stage on which forms of knowledge and reasoning could be performed; this stage was permeated by the uneven power relations between officials and their audience, by cultural assumptions about the forms of reason that officials had to display, and by unspoken norms about what could be said in public and by whom. Both audience and performers at the convention shared the assumption that public knowledge was an illusion that masked the real and dangerous workings of power: As we shall see, at all levels in Mexican society, publicly contradicting official knowledge is considered to be dangerous. This strong cultural association of public knowledge with illusion and danger leads forestry officials themselves to refrain from officially delivering bad news to their superiors: Public knowledge is controlled by senior officials and politicians, while credible knowledge is framed as being produced in more intimate conversations.

This forestry convention may seem remote, far away, and banal: Many of the speeches were certainly quite dull. However, it is in fact precisely in such rather ordinary places and moments that the hard work of performing the state is done. Bureaucratic performances, whether in Mexico or at other places and times, are precisely about producing shared public facts, collective understandings of what the state is, of how technical knowledge is to be produced and assessed, and of the proper role of the public, as witness or critic. This is a coproduction of the political and the technical (Jasanoff 2004), a simultaneous making of things and ways of knowing, ontologies and epistemologies. Understanding the state as the potentially unstable product of a performance that may go awry is potentially radical; it makes us see state-making and related projects of knowledge-making as more dramatic, more interesting and lively, and as potentially unstable and unmade. Official performances tried to make it clear what kind of *thing* the state was: a reasonable, rule-bound, knowledge-collecting object, which spread uniformly into the forests of the state, amassing and acting on statistical knowledge in a reasonably beneficent and impartial fashion. Once we see state and official knowledge-making as potentially unstable, public understandings of the state and collective understandings of forests, fires, and logging come to be much more interesting and powerful. How indigenous people understood forests came to affect not just how they worked in forests, but how they understood and interacted with the state and came to be political and epistemic actors. The textures of these mundane encounters between officials and their audiences turn out to have effects that spiral outward, affecting the stability and credibility of the forest service. In this chapter, I will describe my travels through the public and semipublic worlds of forestry officials, moving from the conference hall in Oaxaca to the offices of officials and foresters in Oaxaca and Mexico City. As we shall see, across these places, officials were haunted by the sense that they did not know what was happening in the forests, that rural people or their subordinates did not do as they were supposed to, and that they, the officials themselves, might be called to account for this gap.

Conventional Knowledge at the Convention: Bureaucratic Authority and the Rhetoric of Numbers

As the convention went on, I became increasingly fascinated by forestry officials' rhetoric, by their recitation of statistics that, they claimed, showed community forestry to be a success. These speeches were often profoundly boring,[2] officials would recite statistical indicators of progress and environmental protection, which, it seemed to me, wouldn't really have motivated or convinced anyone. In a low monotone, the forestry official Aldo Domínguez summarized the "advances of the forestry sector" in the state of Oaxaca over the last few years (field notes, November 8, 2000). He extolled the willing participation of *comuneros* in forest management, and he recited statistics of management plans written, firebreaks created and increased timber production, arguing that these activities had benefited the environment and boosted the economy. His audience listened politely, but in later private conversations with me they were scathingly critical. As one skeptical forestry official told me a few days later:

At the forest forum Ing. Domínguez was putting out all kinds of numbers about reforestation, but he didn't say how many of the seedlings survived. If those numbers were true they would be great. . . . (Luis Mecinas, interview notes, November 29, 2001)

Skepticism about official numbers was omnipresent among officials, foresters, and environmentalists who had attended the convention, but this doubt was only ever expressed in private conversations; there was a clear separation between doubtful public onstage knowledge and credible private offstage knowledge.

There is a general popular skepticism about official rhetoric in Mexico. Even as politicians routinely deploy statistical recitations of success and progress, few people admit to believing these numbers in private conversation. This public doubt and tendency toward disbelief has increased in recent years, as a violent war between the Mexican state and drug traffickers has strengthened public understandings of the state as a mask that conceals a reality of violent and dangerous contests between concealed powers. Certainly, my conversations with foresters, state-level officials, and community leaders gave me the impression that expressing belief in those statistics would have indicated that I was either credulous or a stooge. One day, when driving down from the pine forests of the Sierra Juárez into the dry and dusty valley of Oaxaca, my biologist friend Santiago Perez exploded when he heard Governor Murat on the radio. Murat was declaiming the successes of his administration, the kilometers of roads built, the numbers of schools repaired. Santiago exclaimed angrily to me: "He's just a corrupt politician with his numbers!" This was a private conversation with a friend, but even government officials and foresters expressed a similar disbelief in official statistics in one-on-one conversations that could be framed as intimate.

After the convention, I talked to many professionals who expressed skepticism about Domínguez's statistical rhetoric of success. For members of forest communities, official statistics were less directly criticized because they were less concerned with them. People in Ixtlán were worried about other kinds of numbers: They worried that the numbers in their forest management plan were inaccurate, that their forests might have been mapped out incorrectly, or that timber supplies might be exhausted, but I never heard anyone mention the kinds of statistical indicators of progress that so perturbed officials in Oaxaca. Perhaps community members were simply indifferent to official assertions; as the forester and president of Ixtlán, Leopoldo Ramírez, told me: "The government regulations don't correspond to our reality, we know what's going on in our forests, and those regulations aren't useful to us" (interview notes, July 28, 2000).

State Power and the Danger of Public Secrets

Clearly, neither officials nor their audience had much faith in official statistics: why then the obligatory recitation at the convention, and why did people so carefully avoid criticizing figures they claimed not to believe? People at the conference hall in Oaxaca believed that official statements were likely to be illusions, but they also believed profoundly that puncturing such illusions could be dangerous, that public criticism would result in official reprisals. The dark twin of public acquiescence to official certainty was a widespread belief that official statements were a performance, a mask that concealed the real action—the illegitimate deals whereby officials allocated subsidies or acquiesced to illegal logging. Other researchers have observed that in Mexico official assertions are rarely criticized in public (Lomnitz 1995; Nuijten 2003:133–136), and that audiences often say that officials' public actions and declarations conceal immoral negotiations between powerful actors (Haenn 2005:162–164; Nuijten 1993:200–208). This is a coherent theory of power and knowledge: Real politics takes place behind the scenes, in the dark. Public appearances are no more than a fiction; it is dangerous to openly contradict this fiction, but it is also dangerous to believe in it. Monique Nuijten in particular has discussed this broader understanding of power as being an understanding of an invisible counterstate, an organized mechanism through which resources or official documents can be obtained. While this degree of doubt may seem to be a particular product of Mexican history and present-day circumstances, it resembles the predicament of states around the world, which face a secular decline in their ability to produce legitimate and credible public knowledge (Ezrahi 1990).

In the introduction to this book, I discussed José Chávez Morado's 1940 lithograph, *Nube de Mentiras*. This image vividly represents a cultural imaginary of the danger of official and public knowledge. In this picture, the figure in the foreground is about to step off a cliff, blinded by a storm of newspapers that envelope his head. This is an image of isolation, personal vulnerability, and the danger of believing in what is public. This image refers to political events in the 1940s; it is a polemical warning that workers may be deluded by newspapers. However, this image has a broader resonance: It illustrates the belief I found among officials at all levels that statistics, reports, and documents concealed rather than revealed, that it was dangerous both to contradict official documents and to believe in them. Officials were haunted by the fear that, like the worker in the picture, they might suddenly fall off the cliff, that too blind a belief in official reports from their subordinates might be personally dangerous, that it could hurt their careers. This sense that public knowledge was a performance was sustained by the perception that power was dangerous and that contradiction that was framed as official or public could be personally dangerous. Acceptable criticism and dissent could only expressed in gossip or private conversation, framed as a space of intimate correct knowledge against the impersonal illusions produced by the state. One effect of this understanding of public knowledge as illusion is that credible knowledge is often framed as personal, as coming directly from a friend or a connection. Such retellings of the *real* were frequent in my interviews with officials and foresters, as they would explain the real reason behind public events. What appeared to be violent conflict over community forests or biodiversity protection would be explained as something entirely different, perhaps a conflict over mineral resources or drugs; what appeared to be a forest fire caused by drought was "really" deliberately set by an environmental organization that wished to conceal that it had misspent its budget and allowed forests to be logged.

The danger of contradicting public knowledge and the possible danger of state power were firmly linked in people's minds. Public contradiction was interpreted as a fearful sign of official loss of control, and people firmly believed that such contradiction would be punished, perhaps by loss of subsidies or logging permits. Representatives of the "troublemaking," "rebellious" community of Macuiltianguis have openly contradicted officials on various occasions, and several people told me that the community had been punished for its open rebellion. It was public contradiction that was dangerous: Foot dragging or the failure to follow forestry regulations were less likely to be punished.

Domínguez's statistical recitation in the convention hall in Oaxaca asserted his status as a representative of the Mexican state who could not be publicly contradicted for fear of reprisal. Further, it established the authority of the state over quantitative technical knowledge of the uttermost reaches of forests. This was a rhetoric of transparent knowledge of legible forests, which asymmetrically said nothing about the structure or composition of the state, even as it asserted the state as unified, solid, and coherent. At the same moment that Domínguez's specifically *state* scale of knowledge was asserted, multiple other scales and institutions were invoked[3]: that of the state of Oaxaca, the Mexican state that Dominguez represented, and the global scale initiatives of the World Bank to which the Mexican state responded. Dominguez spoke for these scales in a language of numbers while the material power of these institutions ensured that his audience would remain silent. This performance may not have been believed, but it was highly effective at defining what could be said and by whom. As we shall see, representatives of forest communities could and did complain of neglect by forest service officials, but they did not publicly criticize the factual basis of official statements, nor did they make statistical declarations of their own. Control of forestry statistics by officials also defined the boundaries of the technical and the political, positioning the audience as witnesses to a public knowledge that was protected by shared understandings of the limits and the forms of acceptable criticism.

The reason that official knowledge was so important, that the stakes were so high, and that contradiction was so dangerous was that these performances of knowledge sought to produce the idea of the state and define the realm of the political. Rhetoric of statistical knowledge produced the idea of the forest service as a stable institution that managed forests and helped rural people: The state was that kind of territorially encompassing *thing* that produced and acted on such statistical knowledge. This was then a coproduction of knowledge and politics, where the authority of officials was based on expertise and statistical knowledge, applied impartially and beneficently on nature and society. Domínguez's performance defined the contours of the political and the technical, positioning his audience as witnesses to official technical knowledge, asserting the localness of their knowledge, and preventing them from political criticism of forest policies or technical doubts about official knowledge. Official knowledge then rested on various kinds of silences and complicities: the silences of officials and audiences who avoided touching on embarrassing topics and the silencing that prevented official knowledge from being openly criticized.

Representing Success, Silencing Failure: Rhetoric of Fire and Firewood

A striking feature of Domínguez's statistical panegyric was that he presented incredibly low figures for firewood cutting, failed to mention illegal logging entirely, and only briefly mentioned forest fires and agropastoral burning. In fact, officials generally believed illegal firewood cutting and logging to be at least as large as official timber production, and they worried that recklessly set agropastoral fires were one of the principal causes of forest destruction, but they preferred not to talk about these uncomfortable topics in public. Illegal logging and firewood regulations were barely mentioned and then only to emphasize that it was up to communities to control subsistence firewood cutting. A few days later, Domínguez himself told me that clandestine firewood cutting was many times the official figures. Although he did not acknowledge that he had presented incorrect statistics at the convention, he quickly changed the subject to emphasize that "fire and pests have almost entirely occurred outside of the areas of forest management." Further, he told me he had hoped that the convention would demonstrate the sustainability of community forestry to the hostile state governor and his "misinformed" advisers (interview notes, November 21, 2000). Domínguez had avoided mentioning firewood cutting and burning at the convention because this could have threatened the public representation of sustainable community forestry, undermining his efforts to strengthen SEMARNAP and the PROCYMAF project and weakening alliances with forest communities that were breaking the rules.

Forestry officials at the conference avoided mentioning fire at all, except in sweeping statistical claims that forest fires were almost unknown in "well-managed forests." In private, the same people told me that the vast majority of forest fires were set by rural people and that this was one of the "principal threats" to forests. In an interview a few days later, a senior official emphatically blamed rural people for environmental degradation: "Most of the forest in Oaxaca is degraded because of cutting for agriculture, because of fires, because of insect infestations." Although rural communities were reluctant to authorize logging, degradation continued because "they still cut firewood, and they still do their *rozas* (swiddens)" (interview notes, November 15, 2000). This official imaginary of uncontrolled rural burning was clearly revealed in the work of an educational theater troupe funded by the PROCYMAF project. In this play, a drunken old man who wishes to set fire to the forests because he has matches and it would be fun is dissuaded by his environmentally conscious granddaughter (Gijsbers 2000). Fire, drunkenness, rural disorder, and ignorance are conflated here, illustrating the dark imaginings that officials did not want to talk about at the convention.

Officials expressed sympathy with the poverty that drove poor people to engage in firewood cutting and charcoal burning and told me that it would have been impractical and politically foolhardy to try to control it:

We know that a lot [of charcoal] is being produced, maybe more than the total produced for industrial timber, several times more. The Oaxaca market [where firewood and charcoal were openly sold] is a long term problem, punishing everyone is not the solution. The police or PROFEPA can [forbid sales] for fifteen or thirty days, then political pressures increase and after a month or two the market reopens. (interview notes, January 5, 2001)

These official views of fire and fuelwood cutting were dramatically at odds with the understandings of many of the people in their audience. As we have seen, in the community of Ixtlán, where I spent the most time, controlled burning has ceased and elders there remembered fire as a largely controllable tool for agriculture, scathingly criticizing those who did not burn properly:

One burns to fertilize, one sets fire. Before, the milpa [crop] used to come out perfectly, even though we burned the fertilizer [i.e. vegetation]. First you cut a firebreak around the field, you start at the edges, and when the field is surrounded, you burn from the bottom. . . . (Zenaido Pérez, author interview notes, July 24, 2002)

Many of the people in the audience at the convention came from communities where agropastoral burning continued and would have known of controlled burning by their neighbors, but no one openly suggested that fire was anything other than destructive. Similarly, although people from Ixtlán told me that there was no shortage of firewood for daily use, there was almost no mention of firewood.

Rather than confront intractable and possibly controversial issues, officials like Domínguez sought to present a seamless image of progress and development that was not openly contradicted. This image was stabilized by their audience's collusion and silence; by the presence of representatives of leading communities on the podium; by not mentioning people's concerns over environmental degradation, forest fires, and illegal logging; and by not discussing how logging permits and government subsidies were distributed. The theater of official success was therefore also a place of mistranslation and concealment. It was not that community representatives were quiescent; on the contrary, they sometimes complained bitterly about official neglect and lack of government subsidies and even, on occasion, about complex and confusing regulations. Nevertheless, these criticisms invariably took the form of demands for *more* state attention and *more* resources, rather than criticizing official knowledge or styles of regulation. Enrique Quiroz of the community of Huamuchil bitterly remarked:

If you consulted the [communities and municipalities] instead of changing the law every three or five years maybe the law would work. . . . We don't know the laws, we know books, they are very pretty, but we don't know what is really inside them, and nobody executes them, not even the army. . . . SEMARNAP and PROFEPA are traveling only on paved roads, excuse me for saying so. . . . Sometimes laws have been made which not even SEMARNAP has obeyed, excuse me for saying so. (audiotape recording, November 8, 2000)

Quiroz's claim that SEMARNAP was distant and indifferent was an argument not for less regulation but for *more* contact with the state, for more resources, and for more information. His public protest followed well-understood rhetorical rules: his veiled accusations of corruption and official neglect were acceptable, but he avoided criticizing official knowledge or statistical declarations of success. Further, Quiroz avoided mentioning the fire and firewood regulations that everyone was so carefully tiptoeing around. Neither he nor anyone else ever attacked the fundamental right of the state to regulate how, when, and where they could cut firewood, burn their fields, cut timber, or collect plants and mushrooms.

The texture of encounters between officials and their clients made a difference to the normative and ontological status of official pronouncements. Although people rarely questioned official assertions publicly, officials were less successful in convincing people of the fairness and environmental benefits of industrial forestry. In spite of all official claims, many people believed logging to be corrupt and environmentally degrading. Industrial forestry was also under threat from conservationists who sought to protect forests as a repository of biodiversity rather than as a place to produce industrial timber. The industrial forests that SEMARNAP sought to produce were a very different *kind* from the biodiversity-rich forests imagined by conservationists. Tellingly, the "anti-forestry" governor of Oaxaca distanced himself from industrial forestry by refusing to attend the forestry congress.[4]

Public representations of sustainable industrial forestry were coproduced by officials and audiences who shared a degree of disbelief or doubt, but who also shared an understanding of the proper forms of official knowledge and public dissent. This was a kind of cultural intimacy that paradoxically reasserted the power of the state. Michael Herzfeld uses the term "cultural intimacy" to describe collusive knowing of the "the dirty secrets that provide the basis of lived social experience" (Herzfeld 2005:372). For Herzfeld, cultural intimacy is knowledge shared by members of a nation state about topics too embarrassing to reveal to outsiders. Here I suggest something slightly different: cultural intimacy as shared understandings of public knowledge as illusion, of the state as dangerous, and of what must not be said in public, even with other Mexicans. Knowing what not to say produced various kinds of collusive silences: Collusion in not mentioning embarrassing fires supported industrial forestry and allowed officials to look good before the governor, and collusion in not criticizing figures in which no one believed allowed the coproduction of the state, of politics, and of knowledge. Silence here was both active and productive, as when officials consciously avoided mentioning certain things, and unintentional, as when rhetoric about communities fighting forest fires prevented officials from coming to know of controlled burning.

The central insight here is that official discourses about rural people as reckless burners and destroyers of forest remained unchallenged and, indeed, that they were maintained by practices of not knowing and of silencing on the part of officials and their audiences. Officials in Oaxaca deliberately chose not to talk about firewood cutting and burning by rural people while they firmly believed official discourses on these subjects. Official knowledge of the benefits of industrial forestry also silenced the many alternative knowledges that were present in the room. Environmentalists who thought logging was corrupt and immoral said little, indigenous people who knew about controlled burning said nothing, and activists who thought the forest service was steering subsidies to friends only gossiped about this on the outskirts of the meeting. The conference hall was then a place of silencing and mistranslation, as official knowledge was stabilized by the material and institutional power of the state, by cultural understandings of official performances as dangerous fiction, and by the political alliances among officials, community representatives, and community forests.

These acts of silencing or omission may seem like an accusation against a flawed, insufficiently modern Mexican forestry institution. This is emphatically not my intention: Acts of silencing and suppression of knowledge are a necessary part of the production of public knowledge. Scholars of science and technology studies have long shown how scientists stabilize their representations as facts by destabilizing competing alliances and facts (Callon 1986). For scholars working in this tradition, knowledge is made from an initially fragile network of people, things, and social practices, which is easily destabilized, requiring scientists to engage in skilled rhetoric and constant translation (Latour 1987:103, 110–113). Applying these insights to the study of practices of representation and concealment within the British development agency, DFID,

David Mosse describes how field-level officials represent their activities as emanating from new policies, concealing much of what they actually do from their superiors (Mosse 2005:162–168). Anthropological studies of audit culture in the UK similarly suggest that efforts to assert official transparency for purposes of control produce public representations of assent that bear little relation to local practices, emphasizing that audits are coproduced by the auditors and the audited (Power 1997). Intentional practices of concealment and silencing are clearly present in what are often taken to be centers of modernity and are not necessarily the result of the lack of resources, lack of modernity, incompetence, or corruption in Mexican forestry institutions.

What then is different or distinctive about Mexico? History and culture do make a difference: In the following sections, I will describe the predicament of forestry officials who struggle to build careers within unstable forestry institutions. Returning once again to the image by Morado and stretching its interpretation far beyond the artist's intent, I suggest that officials saw reports and forms as dangerous documents, which could conceal a looming abyss, but which could not be lightly contradicted. Critically, officials at all levels felt that they could not publicly contradict official knowledge even when they knew that it was wrong. Again and again, officials told me how they had to censor reports to their superiors, making sure to report what was welcome and to avoid reporting on failed policies or inappropriate regulations. As we shall see, strategic silence and tactful omission in the face of higher authority were characteristic not only of encounters between officials and rural people in the meeting hall in Oaxaca, but also of relationships between forestry officials and their superiors. Such deference to the forms of official knowledge was not only of local significance: Rather, it affected how knowledge moved through the forest service and how large things such as national forestry statistics were put together.

Knowledge-Making in the Labyrinth: Bureaucratic Careers and the Danger of Reports[5]

During my fieldwork in 1998–2001, I was encountering a relatively new institution, the "Ministry of Environment, Natural Resources and Fisheries" (SEMARNAP). This new institution, founded in 1994, was but the latest in the continual reconfigurations of Mexican forestry institutions: In 2001, a further reorganization moved fisheries to a new ministry and delegated many responsibilities to a national forestry commission,

CONAFOR. I recite this alphabet soup, which is unlikely to interest most people, mainly to emphasize how often reorganizations occurred, how shifting and ephemeral the state was, and how difficult it was to live in such an environment.

When I approached the higher levels of SEMARNAP in Mexico City, I was confused and puzzled by what seemed to me to be a dramatic contradiction between fantastically detailed regulations and the reality of widespread evasion of these regulations in the forests. I wondered about the relationship between officials' public representations, their daily work practices, and their private beliefs. I wondered whether senior forestry officials were ever afflicted with the puzzlement and disorientation that I felt or whether daily life in Mexico City so completely insulated them from life in the countryside that these regulations made sense to them. It was with these questions in mind that I began to talk to senior forestry officials in Mexico City, moving from office to office, asking people about their reflections on what it was like to live and work as a forestry official. I soon realized that my sense of disorientation was shared by many of the officials I talked to. They felt that they didn't know what was happening in the forests or even in offices a few yards away. They too were trying to figure out how SEMARNAP worked and how to make sense of lives and careers in a confusing and unstable environment.

My first impression of the headquarters of the SEMARNAP forestry department was slight shock at how relatively modest this all was. The forestry department offices in the Coyoacán district of Mexico City were a rambling complex of one- and two-story buildings situated in a compound in one corner of the tree nurseries donated to the Mexican state by Miguel Angel de Quevedo in the 1920s. The offices were on a quiet side street dominated by dusty ash trees and *pirules*,[6] somewhat removed from the roar of the city and from the neighboring park, where people ran or walked their dogs and children. I would deposit my driver's license, university ID, or passport with the guards at the front gate (any official looking document would do). In return they gave me a pass, allowing me to wend my way from office to office, asking for directions through a confusing labyrinth of buildings, moving from one official to another, making appointments when they had time to see me, often in the late afternoons, after the long and late lunch hour. It all felt very far from Oaxaca and the forests of the Sierra Juárez, and it was all very confusing. Could these scattered buildings really contain enough people to manage and control the forests of Mexico? In part, the answer was clearly no: How could anyone really control distant forests, from these few offices, using a humble apparatus of forestry regulations and policy documents? Certainly, officials themselves were aware of this tension between dramatic goals and a mundane reality of limited resources, between the need to know what farmers did in distant forests and the reality that they doubted what happened in the office next door. This tension between sweeping regulation and modest resources creates a structural tension in the lives of bureaucrats. On the one hand, as Michael Herzfeld points out, the tension between high ideals and failed policies allows a kind of secular theodicy (Herzfeld 1992:3), which accounts for evil in the world and explains away failure. On the other hand, it left Mexican forestry officials with a profound sense of frustration, doubt, and personal vulnerability. They were supposed to know, but they feared that they did not, and that they would personally be called to account for their failure to know. This is quite different from the forms of collective authorship that allow Pakistani bureaucrats to protect themselves from criticism (Hull 2003). Forestry officials in Mexico struggled to build careers in an atmosphere of doubt, personal vulnerability, and institutional instability.

On paper, SEMARNAP was a formidably large institution with more than 37,000 employees distributed across the states of Mexico, but their responsibilities were even larger, as they were responsible for all aspects of environmental protection, from fisheries to air pollution to controlling toxic waste (see figure 6.1). The number of people directly responsible for forests was much smaller, with perhaps 300 professionally educated foresters for the whole country, although they were supported by a much larger group of secretarial staff, chauffeurs, tree planters, and firefighters. The tension between officials' enormous responsibilities and their very limited ability to fulfill them created a kind of structural paranoia, a sense of being unable to control their lives. The best way to explore this lack of control is to describe the biographies of three forestry officials who worked with official documents and reports, tried to protect forests, and built careers that threaded through new institutions, new forestry laws, and new buildings.

Portraits of Three Functionaries

Antonio Azuela was perhaps in his mid-50s and had already had a distinguished career as a legal sociologist at the national autonomous

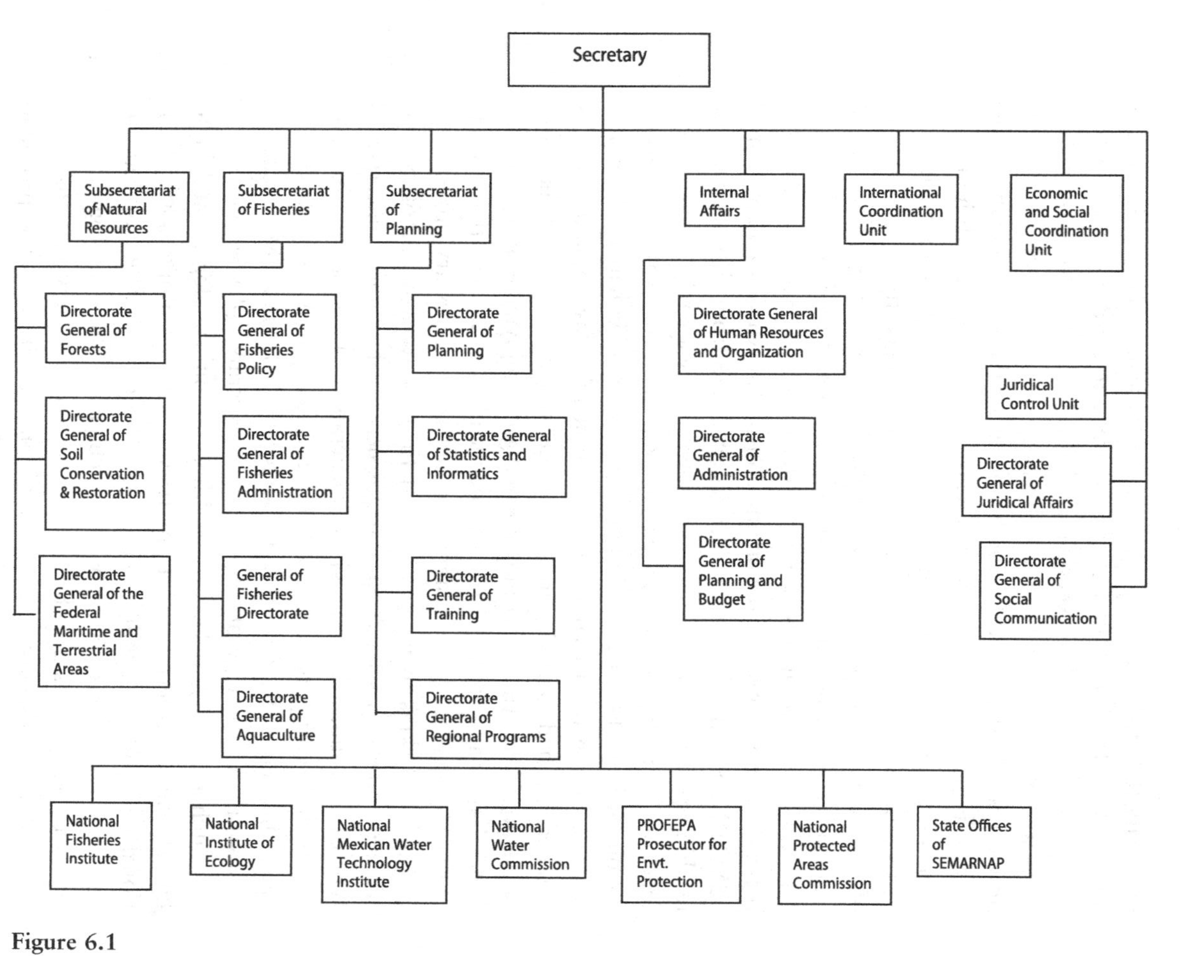

Figure 6.1
Organization chart of SEMARNAP. *Source:* Drawn by author based on SEMARNAP 2001.

university (UNAM)[7]; after six years running the environmental prosecutor's office (PROFEPA), he had just returned to academic life and was happy to reflect on his experiences. I would travel across the urban landscape of Mexico City into the dramatic seared black of the old lava flow on which the UNAM is built and enter the new, recently whitewashed buildings where his offices were. My interviews with him were some of the most entertaining in my fieldwork; he was energetic, animated, and insightful, and our conversations ranged from the daily life of the bureaucrat to meditations on law and history. Antonio was plucked from his academic position to work in PROFEPA by the UNAM biologist Julia Carabias, who had become minister of environment in 1994. This pattern of recruitment is not at all unusual in Mexico: Many senior SEMARNAP officials were academics on leave from their permanent teaching positions, and other government agencies rely on academic/functionary figures.[8] Antonio emphasized that his was an unusual appointment to be an environmental prosecutor: He had never been a senior government official, he wasn't a practicing lawyer, and Carabias and her staff had to work hard to represent him as a lawyer to the office of the president. Antonio communicated great enjoyment in his work as a functionary and in his return to academic life. He was near the pinnacle of the Mexican academic system, and he had the necessary professional and political skills to move back and forth between public office and academic life with relative ease, if not necessarily without uncertainty. When I told him that I thought bureaucrats could be studied like any other ethnic group, he responded enthusiastically and with amusement. Since then he and I have joked about the "Indigenous Movement to Restore the Cosmology of Oppressed Bureaucrats," playing on some of the catch phrases popular among Mexican anthropologists and highlighting the difference between my research and anthropologists' traditional concerns with indigenous people. Antonio said he enjoyed our conversations because they gave him a chance to think through his own recent experiences as a high-level functionary; our conversations had a sense of joint discovery that were illuminated by his own profound knowledge of Mexican law and society.

Aldo Domínguez, who we met briefly at the forestry convention in Oaxaca, occupied a senior position in SEMARNAP in Oaxaca in 1994–2000. He had trained at the forestry University of Chapingo, was recruited into the new SEMARNAP in Oaxaca in 1994, and lost his position in an administrative reshuffle under the new presidential administration in 2001. Domínguez was perhaps in his late 40s when I knew him; sometimes soft spoken and tactful, he could also be brash and direct, describing the "ignorance" of people who disagreed with him. He was rapid speaking and idealistic, working long hours in an office piled high with papers and chairing long meetings of forest councils scattered across the state of Oaxaca. For a mid-level official such as Domínguez, the effort to build a career was more nerve racking than for a senior academic such as Azuela. Domínguez didn't have an academic position to return to and faced great uncertainty about his long-term employment. In his capacity as a mid-ranking state-level official, Domínguez lived the tension between the detailed forestry regulations that emanated from Mexico City and the political realities of life in Oaxaca. This was a difficult balancing act that was made all the more unpleasant by his vulnerability to politically motivated accusations of corruption or incompetence, a vulnerability that was structural to his position as the representative of a state in which few professed to believe.

Francisco Romero was a forester in his early 50s; when I met him, he worked as an administrator at the University of Chapingo. He was a populist, sympathetic to peasants and rural communities and scathingly critical of corrupt foresters and of current SEMARNAP policies, which he saw as being overly conservation oriented. From 1975 to 1985, he had worked in a senior position for a "Forestry Development Section" of the forest service, teaching forest communities how to manage their forests. With the economic crisis of 1985, many extension workers were dismissed, and the department disappeared. In his own words, Romero was "a refugee" who had found a niche within the university system. He was frustrated by the disconnection of the state from the countryside: When I met him, he was trying to reorient forestry education at Chapingo toward more practical and relevant activities.

Initially, I had approached SEMARNAP officials and Chapingo through introductions to Mexican alumni of the Yale forestry school, where I was then a student. I soon realized that beyond their kindness and generosity and perhaps the sense of obligation that my initial introduction might have induced, they were willing to talk to me because they had a burning desire to tell their story. They felt that foresters were misunderstood by society, they felt oppressed by a burdensome bureaucracy that placed impossible demands on them, and they were bitter about the political and moral ascendancy of biodiversity protection. This was made clear to me when I met Saul Monreal Rangel, a young official working in SEMARNAP offices in Coyoacan. Saul was a young and enthusiastic forester who was critical of many aspects of the forestry bureaucracy. Like me he was interested in the history of Mexican forestry, and he shared with me a manuscript he had cowritten subtitled *Chronicle of an Unknown Profession* (Fierros and Monreal Rangel 2000). My conversations with forestry officials like Saul returned again and again to this sense of being misunderstood by society and by their political masters. Foresters felt that the state itself was hostile to their knowledge, leaving them struggling to navigate a confusing and unstable professional world.

Institutional Instability and Personal Uncertainty

At the time and for most of the last century, a key distinction in the working of Mexican government institutions was that between the poorly paid permanent administrative staff, known as the *de base* (base) employees, and the much better paid professionally trained *confianza* (confidence) employees, who had no job security and were usually hired from one administration to the next. With each presidential administration, the vast majority of *confianza* employees would be dismissed. In this atmosphere of ever-present uncertainty about their long-term future, political appointees and senior professional officials try to protect themselves by appointing a team of professionally trained assistants with whom they are personally acquainted and who they feel they can trust not to embarrass or compromise them. More routine administrative work is carried out by poorly paid unionized permanent employees (*empleados de base*).

Traditionally, incoming officials' need to create a team (*equipo*) of trusted associates resulted in the dismissal of thousands of *confianza* officials at the beginning of each presidential administration. In 2004, the Fox administration introduced a new status of permanent career civil servant (*funcionario de carrera*). Officials were allowed up to five years to pass a civil service exam, after which they were to have secure job tenure. This does not seem to have made people feel more secure in their jobs: In interviews in 2008 and 2009, I heard that the five-year term has yet to cover everyone, and that the exam was arbitrary, abstract, and easily manipulated by backstage actors with influence. Some people interpreted this innovation as an effort to entrench PAN affiliated officials within the government and wondered what would happen if a PRI or PRD president were elected.

Certainly, government ministers such as Julia Carabias, who was secretary of SEMARNAP from 1994 to 2000, retain the right to appoint close colleagues who they know and trust, as in her appointment of Antonio Azuela. People like Antonio in turn had to find a team of trusted subordinates. The overall effect was to create a cascading series of dismissals that caused great personal uncertainty. *Confianza* officials have historically had to patch together careers that combine relatively low-paid but more secure academic appointments, periods of public service, and possibly working for industry. The recent changes do no more than introduce another possible uncertainty. Although *confianza* officials live with the constant awareness that they are liable to dismissal at the end of an administration, the actual experience of being summarily fired is unpleasant. As one senior official told me, not long after such an experience:

> The guy who came in [to replace me] didn't even ask me who I was, who is this? They just had us all leave. (Martin Cruz, author interview notes, April 5, 2001)

Being dismissed in this way is made all the more unpleasant by the way that each administration seeks to distance itself from its predecessor by accusing past officials of incompetence or corruption. As the recently dismissed senior official Jose Mares told me of the new Fox administration:

> The new SEMARNAP secretary Victor Lichtinger thinks that all timber trucks have to be controlled . . . that all their documents are false. We all know that there are problems, but who is responsible the [forestry officials] or the forgers? Here it seems [according to Lichtinger] that the forgers are the authorities. (Jose Mares, author interview notes, April 5, 2001)

In Mares's story, the new minister questioned the credibility of past official documents, former officials were vaguely accused of forgery, and these dangerous documents could cause particular careers to suffer. Although official denunciations of former corruption increased after the change in power from the PRI to PAN political parties in 2000,[9] generalized accusations of corruption against past administrations have been a consistent feature in Mexican political life (Morris 1991:83–102).[10] Although she was usually careful to avoid such sweeping accusations, even Julia Carabias occasionally resorted to this kind of generic corruption accusation against her predecessors (Castillo Roman 1996).

Corruption accusations support representations of the new administration as being technocratic and honest, helping build alliances with skeptical publics. Accusations create a conceptual boundary that defines the "new" institution and obscures the continuity of the past with the present, ensuring that past forest policies are not analyzed or evaluated. By blaming past failures on personal corruption rather than systemic political or economic factors, the corruption accusation momentarily bolsters the legitimacy of the new administration at the cost of a long-term decline in the legitimacy of all government institutions. Ignorance about the past is therefore a valued resource that strengthens the current administration. People like Jose Mares well knew that past administrations were not uniformly corrupt, and they complained that simplistic corruption rhetoric caused the lessons of past policies to be systematically ignored so that each administration "started from zero." Highly respected officials such as Mares might be rehired after a few months, preserving the necessary fiction of a clean break with the previous administration. However, their experience of the failures and successes of past policies and legislation was marginalized, making it a personal knowledge that could not be publicly discussed and could not openly enter policy. If we see knowledge in the terms of actor-network theory, as a representation that has a wide network of human and nonhuman allies, we can see how corruption accusations and institutional reorganizations prevented the knowledge of officials such as Mares from building connections with policy statements, institutions, and broader audiences, making it private and marginalizing it from daily practice.

Officials who wished to build a career in this uncertain environment knew that they had to defer to the political and material power of senior politician/functionaries, that documents in particular could be dangerous to their own careers, and that it could be dangerous or inappropriate to openly criticize forest policies and regulations. It is so dangerous to express dissent openly that criticisms of policies are often expressed as anonymous memos that circulate in government offices. Senior officials doubted the truthfulness of reports from their subordinates, and when possible they would try to deal with this problem by hiring trusted subordinates, forming patron–client networks of personal contacts and obligations that brought credible knowledge. The overwhelming power of the presidency and officials' need to scramble for office at the end of each *sexenio* meant that government officials had to be acutely aware of political currents at higher levels and that they tried to maintain patron–client relations and horizontal alliances within other institutions in order to secure employment (Grindle 1977; Lomnitz 1982; Lomnitz, Mayer, and Rees 1983). The proper place to assess such struggles for office was in the informal conversations where people explained the reality that lay behind the official appearances of reform and rationality, explaining who was in, who was out, and why. Writing and documents were too dangerous and too close to daily paperwork practices: This explains the occasional practice of anonymously circulating epigrams and comments within government offices, such as this one by an official who had managed to land a position in the new SEMARNAT in 2001:

I don't know what sustainability is
but we must give it accent. Suddenly it's very kind
because it gives me my subsistence. (personal communication, 2002)[11]

This is a world of profane gossip that both undermines and colludes with the formal authority of bureaucracy and official rhetoric. One of the most eminent practitioners of these cynical and allusive epigrams was the poet and journalist Francisco Liguori (1917–2003), who is reputed to have observed that it was "better to have twenty jobs worth a thousand pesos than one worth twenty thousand," alluding to the relative security of lower level positions. Another mythical narrator of bureaucratic career building is the *Tlacuache*,[12] the poet, novelist, and sometime functionary César Garizurieta (1909–1961), to whom is attributed the remark, "To live outside the budget is to live in error." This proverb is repeated again and again when journalists and officials discuss the struggle for office, and I heard it many times during my research.

Translating and Obscuring within the Bureaucracy

Foresters within SEMARNAP were deeply critical of official environmental policies and felt that political rhetoric prevented their own ecological and professional knowledge from being applied. One senior official, reflecting on thirty years in government service, told me that forest policy was driven by the incorrect environmental beliefs of politicians and their urban audiences, and that senior SEMARNAP officials had to obey their political masters, however much they might disagree with them.

It is not easy for [the public] to form clear ideas, the [senior functionaries] have always answered to the government, so generally the same policies get made, to satisfy in some measure the ideas that [the politicians] have and the same errors are repeated. . . . Politicians say "let's reforest" and money gets spent on this when what is needed is education, training, culture. The national reforestation program (PRONARE) has spent 2 billion pesos and there is not even an area of one hundred hectares where you could come back after five year or six years; all of those trees are dead.[13] (Jose Mares, author interview notes, April 5, 2001)

Mares argued that politicians performed representations of environmental restoration before urban audiences, and that this effectively prevented the failure of tree-planting projects from being reported or considered. He was particularly bitter that a report on the effectiveness of his own program was entirely ignored by the new secretary of environment:

> People say that this government will select the best from every sector; in the forestry sector it didn't happen like that with us. We did an evaluation to see if we achieved our goals, or if not, why not, if it was because of this or because of that . . . and they didn't even read our evaluations. (Jose Mares, author interview notes, April 5, 2001)

Mares and other foresters were particularly incensed by failed tree planting. Perhaps this was because planting and nurturing trees was central to their own professional identity; certainly, foresters like Mares never mentioned controlled burning in the countryside nor did they criticize fire control regulations. Mares' comments were prophetic: In 2008, the failure of mass reforestation projects caused a scandal and the resignation of the head of the national forestry commission in 2009. Clearly, the news that mass reforestation looked good and often failed had not traveled between 2001 and 2009. This recent scandal was produced by a skilled PR campaign by environmental NGOs and by an aggressive parliamentary opposition that demanded the auditing of reforestation projects, indicating a possible shift in the relationship between the state and its publics. In any case, the case of failure to report failed tree plantings until 2009 is similar to the continued failure to report widespread illegal fuelwood cutting and the nonenforcement of fire regulations. Like reports of the death of tree seedlings in reforestation projects, representations of controlled burning and fuelwood cutting failed to travel because they lacked powerful networks of human and nonhuman allies and were silenced by powerful representational alliances between senior functionaries, politicians, urban audiences, and forestry legislation.

Stories about the danger of contradicting official policies continue to circulate. In Oaxaca in 2008, I heard that a forestry official who had criticized failed reforestation projects in what would ordinarily be a safe private conversation with a colleague received a written official reprimand the next day. Such stories demonstrate that mid- and low-level forestry officials believe that criticisms or evidence of failure are unwelcome and may invite reprisals from their superiors. The mirror image of these fears is high-level officials' fear that they do not know what their subordinates are doing. This was probably a realistic concern;

it is so risky for subordinates to express strong opinions that policy suggestions are often made in the form of anonymous memos (*notas informativas*), which are edited and altered as they travel through the system.[14] This was partly because anything written had an almost violent effect:

> Look, if you send something disagreeable, and you sign it, that is really strong. You try and do it over the phone first. People hate to get negative written documents. An administrator hates for you to tell them something critical or bad in writing: his secretary will read it, and he is supposed to be the absolute chief, he has all the power and he doesn't like to lose that image in front of his secretary. Bureaucrats are afraid of the written word. (Antonio Azuela, author interview notes, May 23, 2001)

Insightful officials like Azuela and Carabias knew that that their subordinates' reports could conceal as much as they revealed. This in turn made a government minister like Julia Carabias resort to continual tours of the states where she could see and hear what was going on. In such tours, it is customary for petitioners to present letters and requests directly to the visiting functionary, physically enacting the authority of senior officials and demonstrating the belief of both senior officials and the public that information does *not* move upward through the regular channels of the bureaucracy. Carabias told me that these petitions were a valuable source of extra-official information:

> [you receive] an infinite quantity of complaints, letters, applications. Hundreds. On these work tours people give you an infinite number of documents. . . . This gives you an idea of the local, of how people see the institution, of how your people are behaving . . . you have to be careful because everyone takes advantage to attack each other for political reasons or [they tell] lies, because of historic conflicts, but on the other hand you find out the issues. (Julia Carabias, audiotape interview, July 27, 2001)

As we have already seen in Oaxaca, such encounters between government officials and their audiences are structured by powerful rules about what can be said and who can say it; these petitions and complaints would have been similarly structured while similarly enacting the power of the politician/functionaries to whom they were addressed.

Forestry officials at all levels are haunted by a sense of personal instability and vulnerability and by the fear that they do not know what is going on. A sense of vulnerability is also felt by those who live entirely outside the state, who well know that they will have broken one or another of the myriad regulations that a hostile or malicious official could bring down on them. Officials' sense of fragility and personal dislocation leads them to describe official knowledge as fragile and dislocated also, as in the story told me by the former official Francisco Romero about the 1985 Mexico City earthquake:

> The buildings fell down and sank into the ground and everything that they had archived was gathered up by private contractor and thrown away, perhaps burned or sold for wastepaper. And that was the opportunity for anyone who had been involved in illegal activities to make sure that the paperwork disappeared. (interview notes, April 12, 2000)

Romero's apocalyptic story linked economic crisis, natural disaster, corruption, secrecy and concealment, and the burning of dangerous papers. As I talked to more people, I came to realize that it was not just at moments of crisis such as the 1985 earthquake that files disappeared, people lost their jobs, or institutions were reinvented. Although institutional earthquakes are common,[15] ignorance and loss of knowledge are produced not so much by dramatic institutional collapses as by routine practices of bureaucratic life, by the daily accommodations of junior officials to the forms of official knowledge, and by the accommodations of their superiors to these practices of getting along.

Obscurity, Numbers, and Translations

Senior forestry officials complained about their obligation to produce onerous reports not only on their official work but of their personal wealth as part of an elaborate system of audit aimed at preventing corruption. For all the reports and documents that crossed their desks, they feared that they had no idea what was going on, that a report which crossed their desk and gathered their signature could make them party to the crimes of distant subordinate. Such documents could never be verified, but they could cost them their position. The more reports they saw, the less they believed what they saw:

> They tell you "yes we are going to do it," but then they don't. . . . Their job is to pretend that the machine works as you tell it to. They say that the person who controls an office is the head of the organization, when in reality he doesn't even know if [his subordinates] do their jobs or not, when they don't want to accept what the chief says. (Antonio Azuela, author interview notes, February 21, 2001)

Another former mid-level official argued that all national forestry statistics were a kind of fraud because they were produced by a small number of officials who were extrapolating already inaccurate timber production figures from the 1980s:

The forestry officers have an office in town, but they have no vehicle and [are not] allowed to go out of the forest and see what's really happening. So that database is really from 1985. I have friends in [the ministry of agriculture] who know that they have to adjust their statistics to correspond to what the director says, so he announces a number and then they fiddle the state production numbers to make it. (Francisco Romero, author interview notes, April 12, 2000)

Notice that Romero's criticism of official statistics was based on intimate knowledge provided by "friends", expressing the belief found among officials and ordinary people alike that official statistics are a mask that conceals the real and illegitimate workings of state power.

Regulations, Representations, and Alliances: Fire Control, Firewood Cutting, and Tree Planting

SEMARNAP was an entirely new institution, formed in 1994 by transferring environmental responsibilities from a number of institutions, including the ministries of agriculture (SARH) and environment (SEDUE). This new institution was stabilized by drawing on powerful and enduring framings of reason and expertise; in an atmosphere of widespread doubt and criticism of the state, fighting fires, planting trees, and preventing firewood cutting were attractive and effective ways of establishing SEMARNAP as a source of order and knowledge. New regulations were generated largely in government offices and in negotiations between ministries: Although they might not travel to the provinces, they were used in strategies of institution building. Fire and firewood regulations were both material things and rhetorical statements; they helped establish SEMARNAP as an institution, and they prevented news of rural burning, firewood cutting, and failed tree planting from arriving in Mexico City.

Creating, stabilizing, and organizing this new institution was an enormous task that required senior officials to link laws, regulations, and documents to existing public understandings of nature and the state. In 2001, sitting in a restaurant in Mexico City, Julia Carabias described to me how she had helped produce laws and regulations that defined and sustained the new institution. Carabias and her allies, principally conservationists and members of the community forestry sector, sought to reflect the proper relationship between society and nature within the structure of the new ministry, bringing logging, environmental protection, and biodiversity within the same institution for the first time. The critical 1997 forest law (SEMARNAP 1997b) can be interpreted as a representation of the ideal relationships among the state, nature, and society. Carabias and her allies sought to construct SEMARNAP as a fact, supported by documents, reports, regulations, budgets, and laws. Extending Latour's analysis of scientific facts to bureaucracies (Latour 1990:64–68), we can see how SEMARNAP became a credible representation because it was stabilized by alliances with politicians, the public, with other ministries, and by allies within SEMARNAP itself.

Carabias described the process of consultation and consensus building required to ensure the passage of a new forest law (1997), law of ecological equilibrium (2000), and wildlife law (2000) in the following terms:

> When you try to reform regulations, laws etc., you need very broad based consensus, otherwise nothing prospers. . . . The law of ecological equilibrium (LEGEEPA) was unanimously approved, for the wildlife and the forestry laws we had opposition only from a part of the PRD. (audiotape interview, Mexico City, July 27, 2001)

Carabias was neither a career functionary nor a politician, but a respected biologist and a former critic of government environmental policies. Her recruitment by the Zedillo administration was therefore a sign of the crisis in state legitimacy. She was an innovator and reformer who tried to write popular consultation into the structure of SEMARNAP by creating consultative assemblies on such topics as development and forestry (Blauert and Dietz 2004). Certainly, national-level politicians broadly supported SEMARNAP and the new environment laws, but this unanimity demonstrates that the perspectives of rural people who set fires and cut firewood were largely ignored. The lack of controversy in Mexico City, which Carabias described as demonstrating effective consultation, is precisely an indication of how *little* effective consultation had in fact taken place.

The product of public consultations remained distant from the daily life of field-level officials and rural people. In regional assemblies such as the conference in Oaxaca, the audience expressed little dissent and was more concerned with receiving funds than with expressing their views. In several meetings of regional forest councils that I attended, officials asked for the audience's views on the possible simplification of regulations, but they met with almost universal silence. The unspoken rules dictated that people had little desire to criticize or engage with official regulations. Another reason that new environmental regulations seldom reflected the views of ordinary people was that they were produced through opaque internal and interministerial consultations in Mexico City. The officials who negotiated these regulations were concerned about their logic and aesthetic coherence and about possible conflicts with other ministries, rather than with how the regulations would be applied in forests. This was made clear when I visited an official responsible for drafting three new regulations (Normas Oficiales Mexicanas). In answer to my question as to how these regulations were to be applied, he told me that he was responsible for drafting national regulations: Implementation was the responsibility of state-level officials. The former head of PROFEPA, Antonio Azuela, described this as a general pattern:

> The people who write the law are not the people responsible for applying it, nor even of sending resources so that it is obeyed. . . . Those who write the laws do not respond to the experiences of those who apply them. (interview notes, May 23, 2001)

Fire and Firewood Regulations: Performing a Knowing Institution

Fire and firewood control regulations were dangerous documents that emerged from the consultation processes of 1996 and 1997 (SEMARNAP 1996, 1997c). These regulations were aesthetic and rhetorical objects that spoke to the beliefs of urban audiences and the officials who negotiated them. As we shall see in the following chapter, officials handled such regulations gingerly, deciding which ones to pay attention to and which ones to ignore. Their personal biographies intersected with the biographies of documents as they juggled the regulations' demands for information with their own need to avoid reporting bad news or failure to enforce regulations, balancing the danger to their careers with the danger of filling in forms. Representations of rural burning and unreason were a powerful discourse about nature and society that stabilized forestry institutions and prevented alternative forms of knowledge from traveling to Mexico City. Officials did not passively internalize this official discourse: On the contrary, they made an argumentative use of discourses of rural burning and unreason to do the political and material work of stabilizing forestry institutions (Hajer 1995:56).

Official representations of rural burning and tree cutting were argumentatively performed before urban audiences and other governmental institutions in an effort to build alliances, stabilize representations of SEMARNAP, and manage personal career uncertainty. In an atmosphere of widespread popular distrust and lack of institutional legitimacy, fire fighting, tree planting, and forest protection remain some of the few policies with broad-based popular support. This can be seen in the vast majority of newspaper articles on forest fires, where officials and politicians routinely blame swidden agriculturalists or pastoralists. In a newspaper interview in 2003, the politician Alberto Cárdenas, then head of the National Forestry Commission, argued that 40% of forest fires were caused by "irrational" agropastoral fires, and that farmers who set fires should be deprived of their agricultural subsidies (Gomez Mena 2003). Cárdenas' decision to deploy the time-honored trope of rural unreason reveals the judgment of a skilled political actor who knew the "rules of the game" and the assumptions of his audience of potential supporters.

A similar representation of rural burning and state fire fighting was reluctantly produced by Julia Carabias in Mexican Senate hearings during the catastrophic fire year of 1998 (Senado de la República Mexicana 1998). Under pressure to justify SEMARNAP's actions, Carabias tried to complicate her audience's assumptions about the causes and effects of fire, explaining that not all fires were damaging and that forest fires were not the principal cause of deforestation. Facing a barrage of hostile questions, Carabias eventually conceded that escaped agropastoral fires set in "error" by rural people were a principal cause of the forest fires. Some of her audience criticized fire regulations as overdetailed and unenforced, forcing her to argue that, on the contrary, the regulations were well understood by rural people. It is revealing of the power and political efficacy of representations of official reason and rural unreason that at a moment of institutional crisis and facing an audience who largely blamed rural agriculturalists, Carabias fell back on the official discourse of state knowledge and ignorant rural burning. She told the Senate that well-prepared officials had been overwhelmed by a drought and by the burning practices of small farmers and pastoralists. What was necessary, she concluded, was a transformation of the burning practices of rural people.

Conclusions

In this chapter, I have moved from performances of official knowledge in a forestry convention in Oaxaca to the daily lifeworlds of forestry officials in Mexico City. I have shown how cultural framings of official knowledge as performance and illusion and of documents as dangerous and concealing affect the credibility of official knowledge and the daily documentary practices of officials. By focusing on knowledge as performance and product, I have emphasized how officials and their audiences collaborate in the making of knowledge; this makes knowledge both more valuable and more fragile than formulations of knowledge as a discourse that seamlessly enlists audiences and produces subjectivities. Rather than being imposed on more or less subordinated publics, official knowledge and expertise are coproduced by officials and their often skeptical audiences; these representations may be supported by the institutional power and cognitive authority of the state, but they encounter powerful popular beliefs that can undermine the credibility of official knowledge making. Understandings of the state as dangerous, of official knowledge as performance, and of documents as dangerous caused rural people and environmentalists in Oaxaca to avoid contradicting official performances in which they did not believe or to which they were indifferent. Officials in turn doubted that they could criticize official forms of knowledge in reports to their superiors: They believed that criticism of official knowledge would be ignored or suppressed and that such criticism could hurt their careers. The belief that documents are illusory but dangerous caused officials to engage in careful practices of silence and mistranslation as they avoided reporting that fire and firewood regulations were useless and widely flouted.

Official efforts to stabilize a success story about industrial forestry at the convention in Oaxaca necessarily required the silencing and suppression of alternative forms of knowledge. In these performances, officials braided together multiple forms of power and representation, drawing on ideas about the state, scientific knowledge, and reason; and deploying official discourses about industrial forestry alongside their material power to refuse subsidies, mobilize documents, deny logging permits, and punish recalcitrant communities. This was a making of the technical that also defined the political, but it was not a seamless process. Officials and their audiences were well aware of the danger of politics. Rather than the more or less complete silencing of the political described by Ferguson in Lesotho (Ferguson 1994) or the seamless depoliticizing expertise described by Mitchell in Egypt (Mitchell 2002), this was an expertise where the experts were troubled by the political resistance they were likely to encounter. For Mitchell and Ferguson, the production of the technical is a scandal of knowledge, a way of knowing that improperly silences the political; however, this critique fails to consider that all knowledge-making delimits the technical and suppresses alternative forms of knowledge. What is critical then is that the making of public knowledge is always a public performance that also seeks to define the boundaries of the political. The question is not to reveal the scandal of the suppression of knowledge but to ask how *this* making of the technical and the political happened. What are the roles of publics and experts in making public knowledge? What audiences and forms of knowledge were excluded? How does this affect the texture of politics, official knowledge, and public belief?

Once we see that making ignorance and silencing alternative forms of knowledge is a necessary part of the production of public knowledge, ignorance becomes not a scandal but a political and epistemic resource, and audiences become not the passive recipients of official discourses of development or scientific knowledge but the skeptical audiences who affect the quality and degree of assent in public knowledge and the legitimacy of the state. The textures of encounters between officials and publics affect what kind of a *thing* the state comes to be and how this knowledge is made and witnessed. Alongside the production of public knowledge is a realm of public secrets, complicit alliances, and understandings about what must not be said if officials and their audiences are to collaborate fruitfully. In the conference hall in Oaxaca, official ignorance and practices of silencing had the negative aspect of suppressing criticism, but they also allowed fertile alliances. Officials turned their eyes away from rule breaking in order to enroll forest communities in forest protection and logging; community members legalized their timber production and gained subsidies from officials if they participated in meetings and produced forest management plans. Intimate alliances between officials and their publics were underpinned by shared understandings of the state and its knowledge claims, even as this silencing also prevented officials from registering or reporting how their audience ignored regulations. Cultural intimacy here is the ground for the production of formal official knowledge even as it also undermines it.

Seeing knowledge-making as a kind of theatrical performance highlights the tactical distance between the actor and performance, the collaboration between performer and audience, and the continuous effort required to build and sustain knowledge. The silencing effect of such theatrical performances was recounted to me by Antonio Azuela, describing the moment when he fully realized his transformation from an academic to a functionary whose interactions with subordinates were insulated by power and who magically took on a range of gestures and ways of talking:[16]

My first day at work . . . the union invites me to their annual celebration. I arrive, say a few words, "I'm happy to work with you, our work is important." They say, "We're happy you are with us Mister Prosecutor.". . . When I leave I have to cross the floor where the crowd is. And suddenly, walking out, I realize that lots of people are around me, that people are walking with me, that people are walking backward, that when I stop, they stop. And suddenly, I too begin to walk like a functionary, strong, slow, and decisive, shaking hands on the left and on the right and with cameras flashing. (Antonio Azuela, author interview notes, May 23, 2001)

This vignette shows us a theater of power where subordinates display allegiance but are unlikely to reveal much of what they are thinking. As in the conference hall in Oaxaca, state power stabilized public representations of authority, with highly ritualized statements by Azuela and his respondents. So too, it prevented the "wrong" kind of knowledge from being performed in public. Clearly, this was a ceremonial occasion, where nothing of importance was supposed to be said, but Azuela's description of people walking backward before him vividly evokes the disciplining effects of power, which caused people to display what they thought their superior would wish to see. Far from believing that they knew what happened in the provinces or that their subordinates did what they were told, officials like Azuela knew that subordinates evaded regulations and were unlikely to report unwelcome news. Officials at all levels knew that collusion, concealment, mistranslation, and turning a blind eye were central to the ways that SEMARNAP worked. In the following chapter, we will turn to see how forestry officials in the provinces carried out these tricky practices of transparency and obscurity. For the present, it is worth remembering that these silencing effects are found not only in Mexican forestry bureaucracies, but whenever we attend an official speech, and tacitly accept the boredom of the event, the formal phrases, and the obligation to refrain from interrupting with embarrassing questions. It is out of such interactions that knowledge institutions are built, not only in forests, but in universities, companies, and states around the world.

7

The Acrobatics of Transparency and Obscurity: Forestry Regulations Travel to Oaxaca

State mandates to control forests and protect nature do not travel smoothly through the world. On the contrary, knowledge is continually remade, a practice in translation, rather than an item that travels smoothly from a forestry laboratory or a government office. In the end, a small number of officials and technicians who work in a few office buildings have to build and sustain a web of documents that reaches into distant forests. Returning to Oaxaca from Mexico City, we shall see how forms of official authority and knowledge of forests were sustained by an ecology of relationships among officials, foresters, environmentalists, and forest communities, requiring officials to carry out skilled translations, silencings, and deliberate avoidances. When I was in Oaxaca in 2000–2001, the federal agency responsible for forests was the Ministry of Environment and Natural Resources (SEMARNAP).[1] Most officials worked at the state headquarters in the city of Oaxaca in a modest four-story building in the modern Reforma district, well to the north of the city center. It was a pleasant place to visit: In the late spring, the jacaranda trees that lined many streets would bloom, creating a purple haze over the city. It was often in this neighborhood that I would talk to private foresters (*servicios técnicos*), meeting in cafés or in their offices that were so conveniently near to SEMARNAP headquarters.

Although anthropology of the state has paid considerable attention to the fragmented and contradictory nature of state power, it is only relatively recently that similar attention has been paid to the spatiality of institutions and their projects of place-making, to looking closely at where officials, offices, cars, and management plans are located in space and time (but see Bebbington 2004; Gupta and Sharma 2006; Moore 1998, 2005). All knowledge institutions that seek to reach out into landscapes and societies face a structural distance between sweeping goals and limited means, between the goals of scientific knowledge systems and the reality of weak webs of human and material allies, who may sustain or undermine knowledge practices. As new projects of protecting forests in the name of biodiversity protection or carbon markets become fashionable, it is worth considering just how it is that knowledge about forests is made, sustained, or silenced. As we shall see, practices of silencing were central to the production of official knowledge about forests, not as a kind of defect, but as a direct result of the practice of making official knowledge.

In order to understand the extent to which forestry officials in Mexico or in other places succeed in projects of place-making and authoritative performances of official knowledge, we need to map out the contours of the state, to look at where officials are and what they do when they are at work. In Oaxaca, state efforts to link power to place through forest management plans, maps, and documents were implemented by a fragmented and spatially dispersed institution that had little support from other ministries. Officials' efforts to assert authority and knowledge were haunted by their awareness of institutional weakness, their lack of resources, and their sense that the forests were too remote and too vast to oversee properly. SEMARNAP officials in Oaxaca had to bridge a yawning gap between detailed and unenforceable regulations and local practices, between the enormous scope of their job description and the reality that they had limited resources, too many documents to manage, and not enough time to visit forests. New regulations kept being issued and passed down from Mexico City, but it was impossible to comply with all of them. Officials had to judge how to accommodate regulations that sought to make society and nature transparent, legible, and controllable, while dealing with their own limited resources and likely political opposition. Successful officials did this by a skilled acrobatics that made use of silences, omissions, and concealments, carefully interpreting policy mandates and regulations and deciding whether to act on them or to ignore them discreetly. As we shall see, this was not easy: Some officials were successful in these tricky acrobatics, whereas others slipped and lost their jobs.

Institutional Ecologies in Oaxaca

Officials were supposed to control the 5.1 million hectares of forests in the state (SEMARNAP 1997a:87), divided into 570 municipalities.[2] In practice, the majority of their attention was given to the commercially valuable temperate conifer forests of the Sierra Juárez, which can be seen from the City of Oaxaca itself, as well as the forests of the slightly more distant Sierra Sur. According to official estimates, there were 850,000 hectares of "commercial" conifer forests in the state of Oaxaca, of which 690,000 hectares were under registered management plans that allowed the timber to be cut and sold legally.[3] This meant that most conifer forests (1.8 million hectares) were noncommercial while a further 2.7 million hectares of tropical and dry forest were of no commercial interest and received little attention. The only form of state intervention in these areas was to regulate land use change through the theoretical control of burning, domestic firewood use, and through a system of protected areas (Anta Fonseca et al. 2000:160–161). The vast majority of forest in Oaxaca was owned by communities, but many had no legally recognized logging plan and were largely beyond the reach of the state. Officials classified forest communities according to their level of organization and control of timber extraction (Anta Fonseca and Barrera 2000:30). Around 60 communities were able to log and manage their forests to varying degrees, another 80 or so had management plans but had to trust outside logging companies, and another 150 communities had commercial pine forest but no officially authorized logging and probably rarely encountered a forestry official of any kind.

The total number of SEMARNAP officials was tiny given the vast areas of forest for which they were responsible. Although they represented the Mexican state, they rarely interacted with other state agencies. Unless there was a dramatic emergency, such as a forest fire, officials were on their own. There were approximately forty *empleados de confianza* for the whole state of Oaxaca; these people had degrees in agriculture, forestry, biology or public administration, and were assisted by another two hundred unionized permanent staff in secretarial and lower level administrative positions, as well as by plant nursery workers, chauffeurs, and so on. About half of the *confianza* employees were deployed in regional offices around the state, where they were responsible for extension and fire fighting, leaving about twenty-five senior personnel in the offices in the city of Oaxaca. A typical regional office[4] in the town of Ixtlán had two staff members with university degrees who occupied a modest two-room concrete office with a tin roof on a quiet side street of town. A couple of desks, a back office full of papers, a simple porch, and an official pick up truck with the SEMARNAP logo were the physical manifestations of the state for the tens of thousands of hectares of the Sierra Juárez. The officials in Ixtlán supervised four *empleados de base* who ran a tree nursery and carried out secretarial tasks, with an additional six temporary employees when fire fighting was required.

Although SEMARNAP was responsible for fisheries, protected areas, wildlife, and industrial pollution, there is little industry in Oaxaca, and most officials and employees were concerned with forests. Enforcing forestry regulations was, however, the responsibility of the separate office of the environmental prosecutor, PROFEPA, which had little communication with SEMARNAP. Working from an office in downtown Oaxaca near the old city center, PROFEPA was heroically overburdened, with fourteen inspectors and a staff of forty-nine people spread across the whole of the state of Oaxaca. In 2000–2001, a World Bank-funded forestry project, PROCYMAF, had its offices immediately next door to SEMARNAP headquarters, and a follow-on project, PROCYMAF 2, was still active in 2009. Many PROCYMAF employees were former SEMARNAP officials, and the two institutions collaborated closely. The goal of the PROCYMAF project was to strengthen community forestry institutions and expand the area of forests under management plans so as to expand legal logging and increase the contribution of forestry to the state economy. Most PROCYMAF resources therefore paid for private foresters and other professionals to write forest management plans and studies that supported activities such as fish raising, mushroom production, and producing bottled water.[5]

Daily Paperwork at Headquarters

This brief outline of SEMARNAP's organizational structure suggests how overworked officials were. The office that was responsible for forest management (the subdelegation of natural resources) had five *empleados de confianza* based in the city of Oaxaca. These five overworked officials were responsible for approving forest management plans, environmental impact statements, land use change applications and forestry plantation programs, and for running the government subsidy programs PRODEFOR and PRONARE. As if this were not enough, they were also responsible for issuing the *guías*, the booklets of transport documents that certified the legal origin of forest products, supposedly presenting a documentary snapshot of every cubic meter of timber cut in the state of Oaxaca from the moment it was cut in the forests through its transport and delivery to wood processing centers across Mexico (see figure 7.1). Given this heavy burden of paperwork, they had little time to visit forests and were constantly behind with their work.

This form asks for the name, tax number, and address of the timber shipper and the recipient, as well as for the number, total volume, and species of logs. Such papers are valuable: They may protect traveling timber from police officers' demands for bribes (*mordidas*), and they are one of a numbered series that must be accounted for to SEMARNAP officials. Such forms are also used to collate national timber production statistics (SEMARNAP 1997a).

Before management plans could be approved, they had to move across the desks of many officials, each of whom had to assess the possible risks of adding their signature; plans required supporting documents that were supposed to reassure officials that their signature was appropriate and that they would not suffer for signing. For example, a plan needed a document from the *Secretaría de Reforma Agraria* (Ministry of Agrarian Reform), stating that the person or community submitting the plan was the undisputed legal owner, and for communal lands, that a majority of legally registered commoners had voted to approve the management plan. Although officials tried to avoid authorizing plans where ownership was in dispute, it was almost impossible to do so, and a document that presented a surface appearance of order could conceal dangerous disorder. They knew that documentation alone was inadequate and that they needed personal knowledge of the communities involved in order to judge whether the package of documents on their desk concealed dangerous conflicts that could explode into road closures, hostage taking, violence, or unfavorable newspaper stories that might cost them their jobs.[6]

Management plans also had to be revised for technical rigor: They were exhaustive and expensive to create (averaging more than $M200,000 or around $US20,000 in 2001). They had to contain biodiversity, timber volumes, soils, and socioeconomic data (e.g., SARH and UCODEFO #6 1993; TIASA 1993; CEMASREN 1999). The complexity of the plan preparation process was so great that it required a forty-page instruction booklet (SARH 1994), ensuring that writing such plans remained the expensive and arcane knowledge of forestry officials and *servicios técnicos* (private consulting foresters). Given their overwork, officials in Oaxaca had little option but to accept much of the content of these plans on trust and, when they could, traveling to forests to verify only the volume of pine trees that was mapped and calculated in the plan.

FORMATOS E INSTRUCTIVOS QUE DEBERAN UTILIZAR LOS INTERESADOS PARA DEMOSTRAR LA LEGAL PROCEDENCIA DE LAS MATERIAS PRIMAS FORESTALES

FORMATO: AA-01. AVISO DE APROVECHAMIENTO PARA DEMOSTRAR LA LEGAL PROCEDENCIA DE MADERA EN ROLLO O CON ESCUADRIA

(1) Logo SECRETARIA (visible e invisible logo line) (2) Logo del Gobierno del Estado (visible e invisible, logo line) CON SELLO DE AGUA (3) Número consecutivo del formato (invisible)

(4) DATOS FORMALES DEL REMITENTE	FOLIO AUTORIZADO No. (5)
	FECHA (con letra y número) (6)______ __/__/__ HORA DE EXPEDICION ______
(8) Denominación del Formato: AA-01 AVISO DE APROVECHAMIENTO DE MADERA EN ROLLO O CON ESCUADRIA	FECHA (con letra y número) (7)______ __/__/__ HORA DE VENCIMIENTO ______

INFORMACION SOBRE EL APROVECHAMIENTO O REMITENTE

Nombre del titular o remitente______(9)______ R.F.C.______(10)______
R.F.N.______(11)______ Oficio de autorización No.______(12)______ de fecha______(13)______
Oficio de validación de formatos ______(14)______ de fecha______(15)______
Denominación del predio ______(16)______
Domicilio de donde sale el producto
______(17)______
Municipio______(18)______ Entidad______(19)______
Domicilio fiscal (20)

AUTORIZACION DEL APROVECHAMIENTO

Género y/o producto autorizado ______(21) ______Volumen autorizado ______(22) ______Vigencia ______(23)
______
Anualidad:______(24) ______ de ______ año Tipo de autorización: ______ (25)______

INFORMACION SOBRE EL DESTINATARIO

Nombre______(26) ______ R.F.C.______(27)______
Domicilio de destino de los productos forestales______(28) ______ Población______(29)______
Municipio______(30) ______Entidad______(31)______
Domicilio fiscal______(32)______

INFORMACION SOBRE LA MATERIA PRIMA QUE AMPARA ESTE DOCUMENTO

(33) Número de piezas	(34) Descripción	(35) Volumen y/o peso	(36) Unidad de medida

INFORMACION SOBRE SALDOS

Volumen o peso validado para transportar (37) ______ Fecha (38)	Saldo disponible según formato anterior	(39)
	Cantidad que ampara este formato	(40)
	Saldo que pasa al siguiente formato	(41)

INFORMACION SOBRE EL TRANSPORTE EMPLEADO

Medio de transporte______(42) ______
En el caso de automotores
Marca______(43) ______ Modelo ______(44) ______Tipo______(45) ______
Capacidad______(46) ______ Placas o matrícula______(47) ______
Nombre del propietario del vehículo (48)______

		DESTINATARIO
______ Nombre y firma del chofer (49)	______ Nombre y firma de quien expide (50) ______ Código de identificación (51)	______ Firma de recibido y sello (52)

Obviously the richest information in the management plan . . . concerns trees, what type of species, what number of trees per hectare, what volume per hectare, what treatment[7] they are going to apply, how much they are going to take out, how much they are going to leave, why they are going to take out one quantity, why they are going to leave another. In most cases we check to see if standing timber volumes described in the plan really correspond with what we see in the field. We have had problems with people who are acting irresponsibly, and this makes more work for us, because we have to correct the proposals here many times. There are times when you revise it once, send it back, revise it again and again, so these proposals take a lot of work. (Ipolito German, audiotape interview, November 27, 2000)

In practice, *servicios técnicos* told me that they had neither time nor money to pay for biodiversity inventories so that the ecological, biodiversity, and socioeconomic components of management plans were largely copied from general works and government publications. Officials' evaluations of forest management plans were therefore mainly concerned with making sure that all of the required headings had been addressed rather than with verifying conditions in the forest.[8] Even when corners were cut in this way, writing a management plan was an expensive business; *servicios técnicos* had to organize and supply teams of forestry technicians and assistants to march through the forest, counting trees on tally sheets that were then processed by computer to produce the timber inventories with which SEMARNAP was most concerned.

In the end, officials' principal work was to process documents of many kinds. The interlocutors for these bureaucratic arts were the ecology of *servicios técnicos* and conservation NGOs that surrounded SEMARNAT headquarters in Oaxaca and also became skilled in official paperwork. At somewhat greater distance were the forest communities that had mastered bureaucratic practices on their own account. For *servicios técnicos*, a mastery of both bureaucratic procedure and intimate corruption accusation were required if they were to secure the all important management plan approval that allowed them to mark timber for cutting and to receive payment for their services. For forest communities, *servicios técnicos* fulfilled the vital role of steering their management plans through the bureaucratic approval process, but before even beginning this process, a community required a complete folder (*carpeta básica*) of documents certifying the community's legal existence

Figure 7.1
Form for demonstrating the legal origin of roundwood and sawtimber. *Source:* SEMARNAP (2002).

and its definitive land title. The vast majority of communities either lacked a complete land title and management plan or owned no pine forest. These communities were therefore unable to exploit their forests legally, filled in no documents, and had little or no contact with forestry officials.

Tricky Allies

Faced with this massive burden of paperwork, SEMARNAP officials relied on a cadre of consulting forestry companies to act as intermediaries between the forest service and the forest communities. There were thirty or so of these companies or individuals (*prestadores de servicios técnicos*), ranging in size from a forester with two or three assisting technicians to larger companies that employed up to a dozen people. One manifestation of the intimate relationship between SEMARNAP and the private foresters was the clustering of their offices near SEMARNAP headquarters. Somewhat less constantly involved in forests was an archipelago of environmental NGOs, some of which carried out conservation or development activities in forest communities. It was often difficult to know whether an NGO was a paper organization with no staff and no activities or whether it had a real presence in the countryside. NGO workers gossiped about each other constantly, telling stories about who did real work and who stayed in town. One NGO worker complained that his competitors paid villagers to attend meetings:

> They ought to know a lot about coffee production . . . but there are no results. There's clientilism between them and SEMARNAP. . . . People don't know much about the laws, and the NGO's use this to justify their money (*lana*). Public meetings are for justifiying SEMARNAP and NGO's before the World Bank. (interview notes, December 8, 2000)

Like the *servicios tecnicos*, conservation and development NGOs were intermediaries between a national or an international bureaucracy and rural communities; they used this position to sustain themselves and provide services to outside institutions and to their clients in communities. Broadly speaking, NGOs were populated by biologists and rural development specialists, whereas *servicios técnicos* were run by foresters, many of whom were trained at the national forestry university at Chapingo. In 2000–2001, the difference between environmental NGOs and *servicios tecnicos* was gradually blurring, as environmental NGOs tried to get money for writing forest management plans, and *servicios técnicos* branched out into environmental impact studies and development projects. A final group of organizations in this ecology of institutions were the logging companies, sawmills, and furniture factories that purchased and processed timber for state and national markets. There were about forty community-owned and operated sawmills of varying size spread throughout the state and another forty private sawmills and numerous small furniture factories near the city of Oaxaca.[9] Timber buyers typically purchased from forest communities through intermediaries such as private logging companies and *servicios técnicos*.

This brief outline of the social world of forest conservation and management in Oaxaca illustrates the small number of people involved. Most of the professionals in this world knew each other and had longstanding histories of friendship, rivalry, or outright enmity. Animosities were compounded by competition for financial resources and political connections within communities or with forestry officials. Further, officials and their professional interlocutors were engaged in a covert contest for prestige and professional success. Some private foresters hoped to obtain employment in future administrations, while SEMARNAP officials had to plan for a possible return to industry, academic, or NGO employment if they lost their government jobs.

Corruption Stories and Professional Life

In 2001, I sat in a fashionable café in the pleasant Reforma neighborhood of the city of Oaxaca, sipping organic mountain coffee and talking to a young forester, Gustavo Ruíz. He described to me the world of the *servicios técnicos* who implemented the forestry and conservation policies of the Mexican state and mediated between SEMARNAP and forest communities such as Ixtlán. According to Gustavo, the older foresters were "sharks" who moved through a dark world of corruption and secret dealing; they skillfully wielded accusations of incompetence and financial wrongdoing in order to intimidate, coopt, or coerce their critics and gain control of forest resources.

> Discursively [those guys are the best], they make a great discourse,[10] a well managed discourse, but they are a tremendous mafia. (Gustavo Ruíz, author interview notes, May 14, 2001)

Gustavo was worried that senior foresters could use corruption accusations to undermine young professionals like himself; he trusted his technical abilities, but he doubted that he had mastered the dark art of the corruption accusation. Government officials, foresters, conservationists, and members of forest communities (*comuneros*) all talked of corruption or incompetence, of the secret and dirty dealings that they believed controlled access to forest wealth. This forester had sold communities their own pine seeds for reforestation plantings, that one had fleeced a community by asking for payment in timber, this group of foresters merely wrote management plans but did not reforest, and that group extracted enormous volumes of green timber under the cover of removing trees killed by fire. These stories might appear to be no more than the kind of backbiting and gossip to be expected in a provincial Mexican town, where a small number of foresters, officials, and conservationists had long histories of animosity and competition. However, I gradually began to realize that such accusations were the way in which people undermined the cognitive and epistemic authority of the state and sought to oppose the institutional power of government officials. Officials could monopolize public events with assertions of authoritative official knowledge, but speculative gossip and stories of corruption and incompetence dominated a sphere of intimate conversation, where the "realities" of power were revealed.

For obvious reasons, it was difficult for me to observe people giving and taking bribes, using government budgets to build private houses, or engaging in other corrupt practices. My concern here is not with actual corrupt practices, but with the significance of cultural understandings of corruption and with the political and epistemic work that corruption talk can do.[11] My impression of forestry officials in Oaxaca was that they were idealistic, hard working, and sincere. People did tell me that officials played favorites with foresters and communities, but in spite of the accusations of corruption and incompetence that flew so thick and fast, I almost *never* heard stories of personal enrichment by SEMARNAP officials. Officials were believed to bolster their careers by playing politics, but no one told me that they asked for bribes or payments. The stories I heard about outright bribes were of roadblocks where low-ranking police or soldiers who inspected timber transport documents might demand payment. Stories of corruption and incompetence were occasionally entertaining and often depressing, but they also tried to do political and epistemic work, undermining the authority of official knowledge claims and performances of the state as thing, providing the basis for a counterpolitics that occasionally contributed to the dismissal of officials who had misjudged the strength of their opposition.

Just as senior officials in Mexico City had to manage media representations of their institution in order to build alliances with national publics, officials in Oaxaca had to monitor local newspapers. A newspaper campaign against them could be interpreted by their superiors in Mexico City as evidence that they were burned (*quemados*), that they had aroused such intense opposition that they could no longer work effectively. Historically, officials and politicians have often paid for favorable coverage in national and regional newspapers, so negative media coverage can in fact provide an accurate estimate of possible political opposition. This is no longer so strongly the case at the national level, but in Oaxaca officials told me that newspapers were controlled by nameless local business and political interests who "managed information," mounted media campaigns, and attacked officials. One official told me how hostile newspaper articles had publicized a letter from indigenous communities that demanded that he lose his job for "abuse and corruption":

> Perhaps the journalist wrote this from ignorance, or perhaps someone paid him to make us look bad. You have no idea how much money is paid for this kind of thing. . . . We are obviously affecting strong interests, very strong interests, these accusations [against Alarcon] will favor the illegal timber trade. (Rosendo Alarcon, audiotape interview, July 23, 2001)

These kinds of corruption accusations were potentially effective because of the omnipresent popular belief that official pronouncements of disinterested action and impartial knowledge concealed a reality of illicit and dirty deals. Although official pronouncements could not be contradicted in public because of the material and cultural power of the state, officials had great difficulty in reaching the intimate spheres where corruption accusations were speculatively floated, where the political and epistemic work of undermining official authority was carried out. When I described this kind of speculative conversation to a functionary friend in Mexico City, he instantly recognized it and gave it its proper name, exclaiming, "Oh yes, that's called *medirle el agua a los camotes*," literally, "checking to make sure that the water that is cooking the sweet potatoes has not boiled away." This phrase has something of the flavor of "floating a trial balloon" or of "testing the water temperature," with the additional connotation that this is a potentially dangerous activity that requires constant attention.

Speculative gossip outside of meetings or in other intimate conversations was a way of testing the credibility and political efficacy of a corruption accusation. An accusation had to have strong allies if it was to carry political weight and do work, whether of removing an unpopular official from office or causing a private forester to lose his position as advisor to a forest community. Accusers had to be careful to find political allies; otherwise an accusation could backfire on them. On one occasion, the environmental activist Juan Rosas tentatively suggested to me that many forest management plans in Oaxaca had been illegally and corruptly approved by SEMARNAP officials. This was so contrary to what I had heard from other people that I foolishly expressed surprise and doubt. Rosas hastily withdrew his comments and changed the subject. It was only later that I began to think that he was trying out a suggestion, assessing whether he could make this corruption accusation stick. Perhaps if he had elicited a strong enough response he would have looked for other allies, launched a public campaign, and made political or personal capital. On another occasion, my biologist friend Araceli Lopez hesitated over whether to publicly accuse a respected forester of destructively logging stands of an endangered pine species. She hesitated, weighing the probable furious criticism not only from the forester, concerned, but from the other *servicios técnicos* who might come to the defense of their colleague and make her work in forest communities impossible. In the end, she decided against a public denunciation; the costs of failure were too high, the offense too small, and she lacked the allies to make the accusation stick. A failed corruption accusation could have allowed opponents to label her as *grillo*, someone who carries out politics by gossip, backstage whispering, and corrupt dealings. The term *grillo* is evocative and does political and representational work; it resembles the word for cricket (*grillo*), perhaps comparing accusers to insects who chirp in the night. To successfully brand someone as *grillo* is a way of reinterpreting an accuser as someone engaged in politically motivated gossip, malevolent but insignificant.

Accusations of corruption and incompetence were also used by the *servicios técnicos* and conservation NGOs that brought forestry, conservation, and development to rural communities throughout the state. Community members found it hard to discern which *servicios técnicos* and logging companies could be trusted to pay them for their timber, to build roads, or to carry out promised reforestation activities. In one meeting I attended, community leaders even asked SEMARNAP officials to give them an official list of bad actors, with the hope of avoiding further depredations by logging companies that cut timber and then refused to pay. In this uncertain world, where nothing was what it seemed to be and a successful timber sale could turn into a ripoff, or where an apparently reliable professional could turn out to be in collusion with a logging company, community leaders tried to maintain long-term relationships with *servicios técnicos* and officials. Successful *servicios técnicos* similarly relied on longstanding relationships with community leaders and might deploy their own insinuations of corruption if another forester tried to take over their business. This is what happened to the forester Jose Porrua, who had signed an agreement to manage the forests of a community to the south of Oaxaca. Rival foresters with longstanding personal relationships within the community soon persuaded its leaders that he was incompetent and dishonest, and his contract was canceled. This was the kind of gossip campaign that caused young foresters like Gustavo Ruíz to describe the older foresters as "sharks" or a "mafia." Gustavo felt confident in his professional knowledge, but he was less confident that he could navigate the murky waters in which the older foresters swam. Accusations of corruption took place at the margins of formal events, among the small groups of people who stood at the entrance of meeting halls, in intimate conversations in the offices of foresters and NGOs near SEMARNAP headquarters, or in meetings between foresters and community leaders in small towns scattered across the state.

Acrobatics of Transparency: Fire, Firewood, and Management Plans

Officials juggled with national regulations that sought to produce transparency, with the limited resources and time at their disposal, and with the possibility of local political opposition. This was an acrobatic balancing act in which some were more successful than others. Fire control regulations were curiously the least problematic because they had so little contact with either official or popular practice and could be safely ignored. Juan Soriano, the official responsible for fire fighting in the state of Oaxaca, told me that he had not tried to enforce the new fire control regulation because this would have interfered with collaboration with community fire brigades:

> Often the communities listen but don't apply it. It is a very heavy bureaucratic transaction that we rarely see. . . . It was managed to write a letter[12] (*se logro mandar un oficio*) it was the simplest manner. A letter was sent to the community officials (*bienes comunales*), for them to take official notice of the regulation. (interview notes, Oaxaca, July 7, 2001)

Soriano went on to tell me that officials could not command cooperation and had to work by convincing people (*convencimiento*). His decision to ignore the regulation reflected his skilled judgment about which of his responsibilities his superiors were most concerned with and about the political resistance that aggressively enforcing the fire control regulation could have aroused. When I searched the PROFEPA archives, I could not find a single case of a prosecution for illegal burning in the entire Sierra Juárez between 1994 and 2001. Further, I could not even find a case where a burning application had been filled in. The regulation inhabited an ideal space in Mexico City and moved on paper to villages and towns that it sought to make legible and disciplined, but in fact it made burning practices invisible. Certainly, Soriano did not report his failure to enforce this regulation to his superiors in Oaxaca or Mexico City.

Fire, Firewood, and the Rhetoric of Transparency

The 1997 fire regulation (SEMARNAP 1997c) was a document that asserted legibility and transparency through an attached form. Applicants are supposed to fill in each box on a permission-to-burn form (see figure 7.2), theoretically producing official knowledge and making forests transparent to the state. A farmer or pastoralist who plans to burn is supposed to know how to fill this form in, transposing him or herself to appear to be the kind of person the state allows to burn.[13] Such a person must have an official address, own a piece of land, and know in advance the area of the burn, know the likely date and time of the burn, know when the burn will start and finish, and explain whether the objective of the burn is for agricultural, pastoral, or forestry purposes. This form tried to enlist allies for the state project of knowing what was happening in distant fields and forests. In practice, the form was a material and rhetorical declaration about how rural people *ought* to behave, a statement that the state and its officials represented order and reason, and it produced almost total official ignorance as to where, when, and why people actually burned. A real farmer who was planning to do a burn would be highly unlikely to know in advance the date and time of a burn or the number of participants; such a farmer might not have a simple "agricultural" or pastoral goal in mind; further, he might consider a form a dangerous instrument of official power, better avoided than filled in. The proper subject contemplated by the form was someone who trusted in the beneficence of the state and did not fear official documents; as we have seen, the opposite is often the case.

Like fire control regulations, firewood regulations were ignored by both officials and firewood cutters, but they were more politically problematic. Officially declared fuelwood production in 1997 (SEMARNAP 1997a) was around 1% of the most credible figures (Díaz Jiménez 2000; Masera et al. 1997) and contradicted figures presented within the

Martes 2 de marzo de 1999 DIARIO OFICIAL (Primera Sección) 11

México, Distrito Federal, a los dieciocho días del mes de febrero de mil novecientos noventa y nueve.- La Secretaria de Medio Ambiente, Recursos Naturales y Pesca, **Julia Carabias Lillo**.- Rúbrica.- El Secretario de Agricultura, Ganadería y Desarrollo Rural, **Romárico Arroyo Marroquín**.- Rúbrica.

SECRETARIA DE MEDIO AMBIENTE, RECURSOS NATURALES Y PESCA **SECRETARIA DE AGRICULTURA, GANADERIA Y DESARROLLO RURAL**

NOTIFICACION SOBRE EL USO DEL FUEGO EN TERRENOS FORESTALES Y/O AGROPECUARIOS
NOM-015-SEMARNAP/SAGAR-1997

ANEXO 1
Uso del Fuego en terrenos Forestales y Agropecuarios.

1. Nombre o Razón Social:			
2. Domicilio del solicitante			
Calle y Número:		Colonia, Poblado o Ranchería:	
C.P.	Población:	Municipio o Delegación Política:	Entidad Federativa:
3. Identificación personal del solicitante:			
4. Nombre del Predio y tipo de tenencia:			
5. Ubicación del Predio donde se efectuará la quema:			
6. Superficie a quemar (hectáreas):			
7. Fecha y hora estimada para realizar la quema:			
Fecha en que se hará la quema:	Hora de inicio:	Hora de término:	
8. Objetivo de la Quema:			
Agrícola	Ganadero	Forestal ______	
9. Descripción del tipo de vegetación a quemar:			
Ocochal o acumulación de hojarasca.		Desechos de desmonte (autorizados).	
Residuos de cosecha agrícola.		Desecho de acahual o vegetación secundaria.	
Pastizales en potreros.		Cañaverales.	
Pastos en áreas boscosas.		Otro (especifique)______________	
Desechos de aprovechamiento forestal.			
10. Descripción de las características topográficas del terreno:			
Plano.		Loma.	
Ladera.			
11. No. de personas que participarán en la quema:			
12. Método para realizar la quema:			
Trabajos de preparación para la quema:			

Figure 7.2
Section of the Application for Permission to Burn Form. *Source:* SEMARNAP (1997c).

firewood cutting regulation itself (SEMARNAP 1996). These inaccurate national figures were the direct result of work by provincial and local forestry officials who avoided reporting their inability to enforce fuelwood cutting regulations and submitted figures about the tiny number of legal fuelwood cutting permits to their superiors in Mexico City. These inaccurate national figures ended up being repeated in declarations of statistical authority by officials such as Domínguez. There were good pragmatic reasons to ignore fuelwood and charcoal use in towns like Oaxaca. In 1990, efforts to prevent illegal charcoal sales had caused riots that shut the main market down for two days. As the PROFEPA inspector Luis Barragan told me:

> The truth is that the market is a powder keg. If we go there to inspect, everyone will rise up to stop us, even the person who is selling tomatoes, they will block the roads. . . . In 1990, before PROFEPA, the old forest inspectors went into the market. After two days the governor told them to return the wood they had confiscated. The leaders of the *mercado de abastos* are very powerful and have political connections, and they live from the money that comes from each market stallholder. (interview notes, November 29, 2000)

Barragan was a young and idealistic man who explained that fuelwood markets were tolerated as being customary (*usos y costumbres*[14]). Although he was supposed to inspect trucks laden with fuelwood or charcoal, he was sympathetic with the plight of poor rural people:

> I have seen some very sad cases. We stopped a three ton truck, laden with bags of charcoal, firewood, poles. The different people in the truck said, "Those ten bags are mine, those five are mine, those twenty are mine." There was a woman with a sick child, and the charcoal was to pay money for her to go to the doctor, we could see it was true, the child was there, it was money for the doctor. Then, when you act as inspector, you have to have your criteria (*criterios*) . . . without leaving too much judgment (*criterio*) to the inspector. (interview notes, Oaxaca, November 29, 2000)

Barragan implied that he had allowed the truck to proceed ("You have to have your criteria"), but he immediately contradicted himself, perhaps emphasizing that this was not an act of arbitrary authority or that he had failed to enforce the law. A subordinate was less complimentary. In a long, whispered interview some days later, he told me that PROFEPA was excessively "police-like" (*policiaca*), that it should work to educate rather than to punish.

Whether for reasons of sympathy or political necessity, foresters and forestry officials often had to collude in rule breaking, resulting in an effective silencing of knowledge of burning and firewood cutting. Fire control regulations were simply ignored, and no one filled out the permission to burn form. Fire fighting officials therefore only encountered large and out-of-control fires that, by definition, were no longer controlled agropastoral fires. Official knowledge of controlled burning was made impossible by the articulation of unenforced regulations, state practices of fighting large fires, traditional practices of setting small fires, and the biophysical presence of fires on the rural landscape. The much larger number of small, controlled fires lacked allies and were invisible to officials. The case of firewood cutting was somewhat different, requiring officials to exercise discretion and tactical ignorance. As we have seen, SEMARNAP officials knew of firewood cutting but preferred not to speak of it in public, and they could avoid facing any contradiction unless they were actually engaged in verifying transport documents. It was the *servicios tecnicos* who personally administered forest management plans who faced the greatest contradiction in enforcing the firewood regulations. They had to either turn a blind eye or encourage communities to legalize firewood cutting, all the while worrying that they might be called to account by higher authority. For example, the forester Mauro Ruíz told me that a community he worked in was transporting illegally cut firewood with documentation that carried his signature:

> This was a problem for me as the forester . . . because if PROFEPA [the environmental prosecutor's office] came along, I would be responsible. (interview notes, Oaxaca, January 24, 2001)

I never heard of a forester actually being punished for the misdeeds of community members, but Ruíz clearly feared that this could happen. In fact, the people most likely to suffer official sanctions were *confianza* officials within SEMARNAP itself. When administrations changed, sweeping accusations of corruption or incompetence could cost them their jobs.

Overall, fuelwood cutting was known to officials and was occasionally mentioned in regulations or laws, but it was remote from the daily practices and institutional structures of forest administration in Oaxaca and Mexico City. Officials worked with and touched papers, documents, reports, and budgets. Fuelwood traveled from forests to towns without documents, produced few statistics, and generated little financial support (see figures 7.3 and 7.4 for images of fuelwood and charcoal transportation and sale). Fuelwood entered the hazy world of the half known and the occasionally thought about, a fact that was acknowledged but

Figure 7.3
Fuelwood and charcoal for sale in Oaxaca market.

uninteresting. In interviews with forestry officials in Mexico City, I would routinely raise the question of fuelwood, and just as routinely I encountered a polite lack of interest. The rare exception was Jorge Gomez, an academic who held a senior position in SEMARNAT in 2001:

> As regards charcoal and firewood, Mexican forest policy [on this issue] has been a secret which has been resolved by ignoring it, [no one] has wanted to confront this problem. In 1995 FAO[15] sent us one of the world experts on the subject, someone who knew all the methodologies for carrying out a rigorous study, he trained people here . . . a whole series of people. The results and proposals [of this study] were kept in a drawer, and locked up. (interview notes, May 25, 2001)

Gomez used the familiar political language of secrets, of "real facts" being concealed by locking dangerous documents in a drawer. Far from being a secret, rampant fuelwood cutting has been repeatedly rediscovered over the last century, only to be repeatedly dropped into oblivion and disinterest. As far back as the administration of Quevedo, the forest service attempted to regulate fuelwood exploitation and published studies of charcoal use (Barriguete 1941). Official interest in fuelwood cutting

Figure 7.4
Firewood truck descending from the Sierra Juárez.

and fuelwood plantations resurfaced briefly during the late 1970s (SARH 1979; Jayo Ceniceros 1980). The recent FAO studies to which Gomez referred (Masera et al. 1997) were commissioned by SEMARNAP only to be ignored once again. This constant "rediscovery" of rampant fuelwood cutting reveals the importance of institutional connections and political alliances for the stabilization of knowledge. Firewood has no documents, no budgets, and no allies; regulations slip over collusion and avoidance; and firewood is ignored and unknown to the state. The texture of negotiations between forestry inspectors at distant roadsides turns out to affect the quality of national statistics and the way that people understand statistical authority.

Falling Off the Tightrope

I have used the metaphor of acrobatics to describe the predicament of provincial forestry officials because of the difficult balancing act they had to sustain and because they could so easily make a mistake and fall off. They were supposed to enforce complex regulations that made forests legible to the state, but they also had to know when to produce illegibility and turn a blind eye, which regulations not to enforce and on whom. A misjudgment could result in protests or accusations of incompetence and cost them their jobs. This is what happened to Domínguez, who irritated foresters by asking for accurate management plans but whose cardinal offense was probably a failure to maintain a public appearance of civility and assent. He was massively overworked and oppressed by his many obligations: His office was piled high with the voluminous management plans and cutting reports that he was supposed to certify but that he had little means of verifying. Foresters and community members generally agreed that he was honest; in an environment of omnipresent corruption accusations, it was telling that no one accused him of corruption. People did accuse him of being "unrealistic,": of standing too much on the letter of the law, and perhaps more seriously, some interpreted his efforts to improve management plans as a mask that concealed a policy of rewarding allies and punishing enemies. Domínguez told me that he struggled to get foresters to include the most basic elements of truth in their reports, as when cutting plans reported ten times the timber that was really there or when they proposed logging in areas with no pine trees at all. Needless to say, *servicios técnicos* resented this additional work and expense and complained bitterly about his "unrealistic expectations."

In spite of this disquiet among *servicios técnicos*, the first public sign that Domínguez had made a misjudgment occurred during a tour by the new head of SEMARNAP, Victor Lichtinger, in the spring of 2001. The environmental activist Juan Rosas gleefully described to me how leaders from one community had openly criticized Domínguez for approving a management plan on contested land, later repeating these accusations in a regional forest council (fieldnotes, Oaxaca, May 1–2, 2001). Such events were typically marked by the kind of acquiescence and silence that I have described in the forest forum in Oaxaca, so this public dissent was highly unusual. People from the community concerned told me their forest was being logged by a corrupt elite in neighboring communities, in collusion with Domínguez and other SEMARNAP officials. As both Rosas and I knew, public events like the forest council were supposed to function smoothly, and community leaders usually refrained from public criticisms for fear of reprisals. Rosas therefore interpreted public criticism as a catastrophic failure and a sign of Domínguez's weakness. Events were to prove Rosas correct.

From Domínguez's point of view, accusations of favoritism or corruption were cruelly unjust. He told me that he had tried to enforce rules impartially and distribute subsidies fairly, and he even expressed frustration that community members were not more vocal in public events. In 2001 he was dismissed, officially because of budget cuts and a reorganization but, according to gossip, because his superior had cut him loose, saying, "He is not one of my own." This story was gossip, but it reflected a perception that Domínguez was a political liability because of his failure to maintain a seamless public representation of success.

I well remember his bitterness at our final meeting, when I asked him to reflect on his time in office. He spoke specifically about the public denunciation that had so gravely weakened his position:

> The foresters misinform the landowners, and then get them to come and protest at SEMARNAP headquarters. They are a terrible *grilla*. . . . What bothers me is the continual coming and going. I get tired, that's the only thing that I feel. In Oaxaca governors have relationships with powerful groups who don't want development, they would rather have a vote in their favor. (interview notes, July 4, 2001)

Domínguez had failed to maintain alliances that would sustain representations of his own success that were credible to his superiors; further, he had misjudged how to apply regulations on *servicios técnicos*, how to decide what to make legible to the state, and what to ignore. If a successful public fact is the building of a network of alliances between humans and nonhumans (Latour 1987, 1990), Domínguez's network turned out to be insufficiently solid. Domínguez and his superiors had tried to build a solid representational alliance at the forestry convention in Oaxaca. Although they succeeded in momentarily silencing opposition from conservation activists who opposed logging, Domínguez's representations of competence and success were subverted by the counterpolitics of accusation and innuendo that resurfaced as open dissent before the new head of SEMARNAT in 2001. Officials such as Domínguez faced a painful dilemma: They might be able to produce apparently authoritative representations of official knowledge, but in the long run such performances lacked credibility and broad support. Many people preferred alternative explanations that emphasized corruption and environmental degradation and that provided them with political traction against the epistemic and institutional authority of the state.

State-Making and Knowledge

In this and the preceding chapter, I have argued that various forms of official ignorance—of silencing, mistranslation, and collusion—are part and parcel of the projects of making states and of making knowledge.

Producing ignorance can be as important as producing knowledge, and it is a form of power rather than a defect in the operation of the state. I do not argue that the Mexican state tries to make society and nature legible and then fails to do so because of corruption and incompetence, nor do I think that this is merely a case of "muddling through" where officials do their best with limited resources. On the contrary, producing official legibility by its nature requires officials to carry out skilled practices of making illegibility (Scott 1998:11–18).

Unlike the scientists described by Bruno Latour and other ethnographers of laboratory life, Mexican forestry officials could deploy their material and symbolic power as representatives of a state, a fearful and glamorous power that could reward assent and punish open contradiction. Shared cultural understandings of the state caused officials and their audiences to see official performances as a kind of theater that masked the reality of power. People also believed that contradicting the performance was much more serious than ignoring a regulation after the event. Contradicting the public secret of official disinterest, progress and development invited retaliation, whereas foot dragging or rule breaking could be ignored. It was this mask of formal assent that made the counterpolitics of gossip and innuendo so powerful and that allowed accusations of corruption, favoritism, and incompetence to flourish and do political work. This points to the cognitive and epistemic effects of political culture: Histories of Mexican state-making sharply limited officials' ability to construct the state as the kind of thing that knew forests and regulated nature and society. Officials also failed to construct themselves as representatives of impartial technical knowledge, making them vulnerable to stories of corruption that undermined their technical and moral authority.

The frustrations of Mexican forestry officials and rural people reveal that knowledge-making requires continuous cultural, political, and material work. Knowledge is not an object that travels from laboratories to publics (Callon 1986), ideas do not naturally flow through the world (Tsing 2005, 89–112), nor are they easily imposed by officials on their audiences. As any frustrated professor knows, knowledge is fragile and constantly in translation, sustained by practices that draw on material objects and enlist audiences. Knowledge of fire and firewood cutting failed to travel to Mexico City because it lacked powerful allies that could enlist senior officials, politicians, and urban audiences. Fire and firewood regulations were material documents that were important in establishing the state as the source of order before its urban audiences.

Knowledge of controlled burning and firewood cutting, of how and where people cut trees or carried out controlled burns, was silenced by official knowledge of industrial forestry and development and of rural people as reckless and destructive. Where officials and their audiences in Oaxaca colluded in bland assent, some kinds of knowledge failed to travel, to be stabilized, or to flourish, whereas official knowledge received only shaky support. Behind official assertions and bland assent lay a failure to respond to the moral imaginations of the audience, to their concerns about justice or environmental degradation. These concerns resurfaced in the politically efficacious stories about corruption, incompetence, and environmental degradation that caused Domínguez to lose his job, in the repeated reorganizations that left officials like Mares so angry, or in Romero's stories of fictitious statistics, earthquakes, and disappearing archives.

Mexican forestry officials' intentional and unintentional practices of making knowledge and ignorance contradict assumptions about official knowledge, whether as a power-laden discourse (Escobar 1995; Ferguson 1994) or the result of a project of legibility (Scott 1998). Drawing on science and technology studies and recent ethnographies of bureaucracy, I have highlighted the ability of officials and their audiences to partially ignore environmental discourses and to collaborate in making knowledge and ignorance. Bureaucratic knowledge-making is profoundly context dependent, requiring skillful intentional practices such as those required to fill in and or ignore a form (Brenneis 2006). These practices are only partially constrained by official discourses or projects of legibility, because knowing how to make legibility precisely requires a certain distance from the form, an ability to ask, "What fits here?" and "Would it be better to ignore this one?" This freedom of action has the effect of drastically undermining the ability of the states to know things and to enlist people and things into projects of making transparent knowledge.

The texture of official engagements with locals has profoundly extra-local effects: Where audiences collude with officials in evading official forms of knowledge-making, a symmetrical form of official ignorance is created. No forms about fire or firewood travel through the bureaucracy, and no knowledge about fire and firewood is created. Once knowledge is seen as always and inevitably linked to wider publics, even when it appears to be most distant from them (as when buried in the laboratory) (Wynne 2005), the texture of public beliefs comes to be important not because it is "local" or because these publics must be recipients of traveling scientific knowledge. Rather, local beliefs affect what institutions can know, what papers will travel, what reports will circulate, and what knowledge officials can have. Cultures of knowledge, framings of expertise, and beliefs about the state are therefore critical in affecting how knowledge becomes anchored in institutional practices, official ways of knowing, and popular belief (or doubt).

Where official knowledge-making does not find local allies and interlocutors, it slips smoothly across assent; this slippery transaction produces a symmetrical absence within the forest service itself. Fire and firewood regulations leave little record, have no effect on institutional structure and practice, and produce official ignorance as to where and why people burn fields and cut firewood. This yields the profoundly counterintuitive conclusion that in Mexico official knowledge of forests can be the product not of authoritative state institutions but of the ability of relatively weak officials to build alliances with powerful indigenous communities. Official knowledge proceeds not by imposition alone but by entanglement, mistranslation, and concealment, as officials seek to engage their audiences in public knowledge-making. Socially accepted knowledge of Mexican forests is the product of "civic epistemologies" (Jasanoff 2005:247–271), where the beliefs and practices of citizens affect whether official environmental discourses and regulations are entangled with popular practice and assent or slip smoothly across silence and disbelief. Knowledge about the forests of Oaxaca was not imposed on rural people by powerful officials, nor did officials succeed in "reading" the forests. On the contrary, knowledge was coproduced by performer and audience according to the well-understood rules of Mexican political culture; there was a certain intimacy to this knowledge-making, as performer and audience agreed not to disagree, not to mention embarrassing facts. I will turn to this coproduction of knowledge in the following chapter, where we will see how the political, symbolic, and material clout of communities such as Ixtlán allowed officials and rural people to collaborate in producing shared knowledge.

8

Working the Indigenous Industrial

Let us return for one last time to the community of Ixtlán to look at how indigenous people and their community institutions entangle distant pine forests with the content of national forestry statistics and the stability of federal forestry institutions. A central argument of this book is that what appear to be insignificant details are not necessarily so, that national statistics and official knowledge are coproduced between states and publics rather than simply imposed by fiat. If this is so, the textures of these moments of coproduction are all important and not incidental at all. Ethnographic details can tell us not something additional and entertaining, but something critically important for the stability and legitimacy of official knowledge and of state institutions. These moments where knowledge is coproduced are also moments that coproduce ignorance and nonknowledge of various kinds, not as pathologies or failures but as part of what sustains knowledge itself. Knowing and not knowing are woven into the making of knowledge, not only in government offices and meetings with officials, but in the practices of working in forests.

So far in this book, I have described how forestry science travels from universities and government offices to distant forests and remote indigenous communities. In this last chapter, I will describe a little of what it is like to mark and cut down trees, to conduct a meeting in an indigenous community, and to try to form an alliance between an indigenous community and a federal bureaucracy. Given the nature of historical evidence, I could not be a fly on the wall (or perhaps a fly on a tree?) in those past moments when forestry traveled into forests, when loggers and foresters negotiated how to know what forests were, how to cut down, measure, or count trees, and what this meant for the legitimacy of the state or of their municipality. These small-scale details about indigenous community meetings and encounters between forestry technicians and loggers come to affect the stability of federal forestry institutions that yearn for stable community interlocutors. Official knowledge of what happens in forests turns out, in this case, to depend on the mutual vulnerabilities of forestry officials and indigenous community leaders. This kind of mutual vulnerability is present at other levels also, often with very large consequences, as when forestry technicians and loggers negotiate where trees will be cut and how they will be counted, so that national forestry statistics are produced not only by decisions in Mexico City but by mundane practices of paperwork in distant forests. Knowing and not knowing live side by side in the judgments of forestry technicians who decide what to record and what to ignore, what counts as a log or a marked tree, and which loggers it is best not to be too firm with. As in past moments of encounter between the state and indigenous communities, these remote forests turn out to be not so very remote after all. These are places where forest workers continually try to entangle other places, institutions, and forms of knowledge into contests over managing forests, dividing up profits, and allying or opposing state forestry institutions.

Traveling to Ixtlán

When I first made the two-hour bus ride from the city of Oaxaca up to Ixtlán in the summer of 1998, I had no idea what to expect. I had been told that Ixtlán was a "model forest community," that it was high in the mountains, that it was indigenous, and that it had much cloud forest. As the bus wound its way up into the hills and we traveled first through scrubby oak and oak pine forests and then through dense pine forest, I kept hoping for the mystery of a town set deep among tall trees, with ever-present mists and hanging mosses. Reality was more prosaic: The final sections of the highway went through old fields and thorn scrub, and the pine forests were a dark green mass on the mountains above town. The bus dropped me off at a PEMEX gas station on the side of the highway. From there I walked up into town, always hoping for the exotic but seeing little evidence of it. Most buildings were one- and two-story concrete houses with tin roofs, almost always surrounded by gardens planted with fruit trees—apricots or quinces, apples or medlars. Some houses were simple structures built from unfinished saw mill off-cuts; a few older adobe buildings with tiled roofs clustered around the central town plaza, and there were also many substantial two-story concrete structures built by returning immigrants (see figure 8.1). It was in the municipal palace on this plaza (see figure

Figure 8.1
A view of the town of Ixtlán de Juárez.

8.2) that I had met the *presidente de bienes comunales* (president of community property) from whom I secured permission to carry out my research.

Ixtlán is a small country town of around 2,000 people, but for its size it is notably full of professionals and administrators. This is because it is the administrative center for multiple levels of government, each with its own office and archives, officials and petitioners. The town is the center of a district, of a municipality,[1] and of a community; offices of federal, state, municipal, and community government are housed in the *palacio municipal* that dominates the town's central plaza. Other government offices include a federal court of first instance, a civil

Figure 8.2
The town hall and plaza of Ixtlán.

registry, offices of the Ministry of Interior (*Ministerio Publico*), of federal transport officials, and, at that time, a contingent of soldiers and army officers who were installed after the 1994 Zapatista rebellion in Chiapas. In addition, Ixtlán has a federal prison, a new hospital and library, several schools, and a bank. On any weekday, the plaza in front of the town hall was populated by at least a dozen people waiting to do business at one or another of the government offices. On the other side of the square, a two-story arcade contained a row of offices: A couple of lawyers, a photocopy service, several medical offices, and the office of a parliamentary deputy offered an array of bureaucratic and political connections. The community forestry business, known as the UCFAS,[2] was a bureaucracy all of its own, employing about 200 people, spread between saw mill and furniture factory workers, the operators of a high-tech drying kiln, and the logging truck drivers and tree cutters who worked in the forests far above town (see figure 8.3). Finally, there were the technicians who administered the community forest management plan and with whom I spent time in the forests.

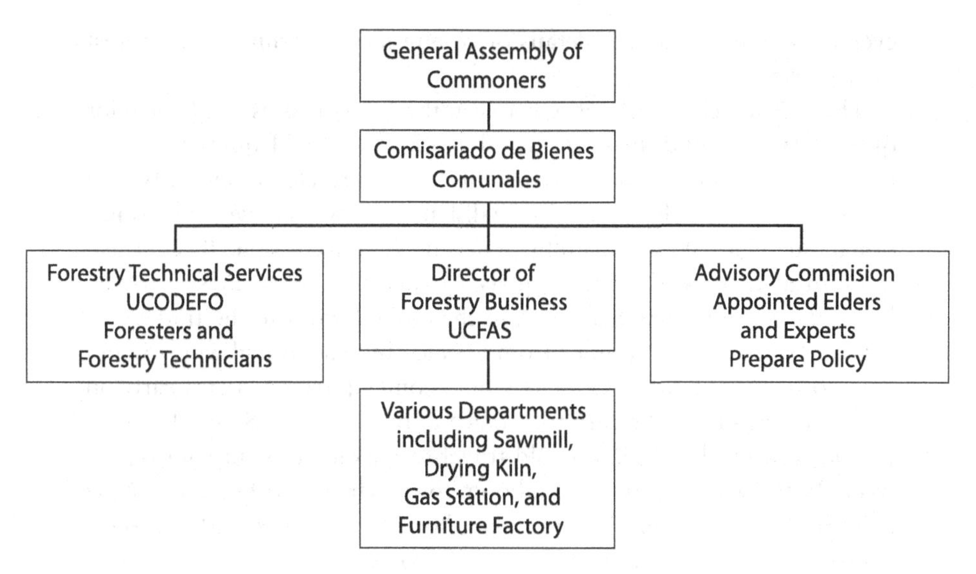

Figure 8.3
Organization of forest industry in Ixtlán de Juárez.

The modernity and sophistication of Ixtlán were brought home to me on my first night in town, when I stayed in a new two-story brick and concrete house built by an entrepreneurial school teacher and merchant. When I turned on the television in my room, hoping to alleviate the first rush of loneliness and the sense of let down at how unexotic everything looked, I found that I could only watch whatever my landlord and his family were watching downstairs. They were fascinated by U.S. Professional Golf Association tournaments, so that my memories of my first weeks of fieldwork are of a mixture of loneliness, of learning to find my way round town, and of images of manicured golf greens and genteelly clapping crowds. I never did talk to my landlord about golf, but as I talked to more people, I came to see that, like me, many of them continually compared Ixtlán to images of other places. Far from being the remote indigenous village of my hopeful initial imaginings, Ixtlán was notably well connected to the outside world, both through technological means and through people's experiences of travel, immigration, and return. I began to notice satellite dishes; in 2002, the first internet café came to town, funded by an *Ixtleco*,[3] who had done well as an immigrant in the United States. People in Ixtlán have relatives living in Oaxaca, Mexico City, and the United States and are constantly aware of the wider world, even as they negotiate the details of community government and forest management.

There was a clear difference between the "urban" dress and behavior of most *Ixtlecos* and the more "rural" look of visitors from the remoter towns of the Sierra Juárez. *Ixtlecos* usually wore closed shoes, button down shirts or tee shirts, and baseball hats; few women wore the traditional blue/black shawls worn by mountain Zapotec people. By contrast, more rural visitors still wore sandals, carried their belongings in woven bags, often wore straw hats, and, in the case of women, the traditional shawls. Ixtlán is only an hour from Oaxaca by federal highway 75, and many officials and office workers would come up from Oaxaca early on Monday morning and return to Oaxaca at the weekend. Similarly, many people who lived in Ixtlán would travel to Oaxaca to study during the week. Ixtlán was something of a boom town when I worked there. New concrete buildings were being put up, both by returning Ixtlán professionals who liked the calm and the good climate, as well as by immigrants who had returned from the United States and wanted to settle down. In 2005, a new university was built between Ixtlán and its neighbor, Guelatao, creating an influx of students, university professors, and other outsiders.

When I was in Ixtlán in 2000–2001, community members were already worried about the influx of outsiders, arguing that not just anyone should be allowed to settle in town, fearful that the town's commercial success would result in outsiders buying up communal lands and progressively displacing *comuneros* "to the hills," away from the center of town. Relationships between Ixtlán and neighboring communities are not necessarily good: Many are administratively subject to Ixtlán, and the community's large forest is partly a reflection of the community's past political success, sometimes at the expense of its neighbors. At present, the community logging business employs outsiders as relatively poorly paid saw mill workers and from time to time purchases timber from nearby communities that cannot do their own logging. Given the history of logging under FAPATUX and the much older history of conflicts, this relationship is inevitably open to distrust and accusations of exploitation.

Making Knowledge by Appealing to Other Places

People in Ixtlán constantly talked about their relationship to the outside world, comparing their town to other places, their knowledge to other kinds of knowledge, and the health of their forests to environments elsewhere. These appeals to other places were used in community politics; appeals to other places could stabilize social relationships and technical knowledge in town, or they could be used to criticize official knowledge. The most surprising appeal was made to me be by Jose Pozo, an elderly comunero who had spent many years in the forest working as a logger.

Pozo was old and angry; his voice shook with rage as he told me how he had been denied senior community offices (*cargos*) because of jealousy and because he was illiterate. He told me that he received no pension from the community forestry business, that he had been unable to secure a job for one of his sons, and that community foresters disregarded his views. He was bitterly critical of community forestry institutions, of fellow community members, and of the technical details of tree planting and forest management. Although he shared corporate ownership of the community forests, Pozo was politically and spatially marginal. Unlike the majority of *comuneros* who had been able to build substantial concrete structures, he lived in a flimsy wooden house on the upper edge of town, surrounded by apricot and peach trees. On our second meeting, he astonished me when he launched into a diatribe that combined an appeal to the charisma of Nazi political power and scientific forestry with his own ecological knowledge and experience of community politics.

> My opinion of the *servicios técnicos* [the community foresters] is that they have things backwards, that they don't work well. I read in a magazine about Rodolfo Hitler in Germany, that he had a good system of managing the forest. That Rodolfo Hitler made himself lively, he got his people together and he organized them to exploit the forest and he did work well. First, because he knew how to control his people; he would have assemblies [*asambleas* i.e. community meetings] in order to come to an agreement about how to work the forests. Then they began to divide up the parcels of lands, by *paradas* [the working team of a logger and his assistant]. Little by little they began to realize, the more they worked there, that the work itself shows one how to work. But there he [Rodolfo] didn't use people to reforest (i.e. replant seedlings), he made the people responsible clear cut and burn, the seed fell by itself and there you get pine. (Jose Pozo, author interview notes, July 25, 2002)

For a long time, I was not sure how to think about this conversation. Certainly, Pozo was unclear about who Adolf Hitler was; somewhere, somehow he had read a pamphlet or magazine that described Nazi era forestry,[4] but I didn't want to think about this as an example of lack of knowledge or rural inauthenticity. Gradually, I came to realize that this kind of story, with its appeal to the charisma of outside power and knowledge, was a typical strategy in Ixtlán. People constantly invoked distant places and events in their discussions of community politics, articulating the forests of the Sierra Juárez with outside institutions, with other places and forms of knowledge, as when the mayor explained to me:

People say that there is no more water, but I don't agree completely, because here, where we have taken care of the forests, it carries on raining, as here [in Ixtlán] where the rains have been good. On a world level the water is finishing, on a world level, but not here, because we have cared for our forests, maybe in other places the water is finishing, but not here. (Graciano Torres, author interview notes, July 25, 2002)

In this case, the mayor was using ideas of global climate change and environmental degradation in distant places so as to claim the virtue and technical skill of the community, strengthening the legitimacy of community forestry.

Like the mayor, Pozo's speech skillfully combined the mundane details of logging practices with the charisma of distant events; he talked of Hitler's ability to bring together a community assembly and apply science, of how he divided the forest up into sections that could be logged by the two-man logging teams. Pozo emphasized the importance of fire in pine regeneration and the uselessness of reforesting with tree seedlings from the community nursery:

And Rodolfo Hitler's *servicios tecnicos* would walk in the forest and it worked well for them. We have a lot of tropical timber and vines, we should cut the biggest trees that are standing, leave the *rozo* (i.e. the dry vegetation) and burn. We used to plant the corn for a year or two, afterwards we go off in another direction and by itself the land will reforest, there is no need to plant, to reforest. Easy, easy, the wind will take care of spreading the seed. The people in charge now look for people to reforest, but they don't know that with burning . . . they don't know that with burning the seed falls by itself, how it is born all by itself. And they pay people to plant little trees, that's what the people in charge now do. (Jose Pozo, author interview notes, July 25, 2002)

Here Pozo used the concepts of swidden agriculture to criticize contemporary forest management. For Pozo, nature was powerful, forest regeneration was inevitable, and planting seedlings was a waste of effort:

And that Rodolfo Hitler, he had it organized, and people saw how they were situated [i.e. how well off they were], hugging a pile of money. Afterwards people went too far, they committed abuses against priests and nuns, they killed them and people got angry with him, they wanted to kill him. People got together in an *asamblea,* they arrived and talked, but Hitler realized what was happening.

Some people say that they killed him, others that he hid beneath the sea, that he survived and that he is still alive. Imagine how much science he had, don't believe that these are lies. (Jose Pozo, author interview notes, July 25, 2002)

Here Pozo combined rumors of the Holocaust with the earlier Mexican experience of the Cristero rebellion of 1926–1929, when rural Catholic insurgents fought against the anticlericalism of the Mexican state. Hitler was gone, perhaps hiding under the sea, because he had lost control of an assembly and committed abuses against priests and nuns, but Pozo could use Hitler's prestige to call into question the technical knowledge of community foresters.

Several threads ran through Pozo's speech: concern and knowledge about the details of forest ecology, familiarity with processes of community decision making and office holding, and criticism of the status and technical knowledge of community foresters. Pozo felt that they disregarded him because of his lack of education and status. By appealing to the outside world in order to argue about logging practices, Pozo sought to weave skeins that might destabilize community forestry and change tree-planting techniques.

Community and Indigeneity

As Raymond Williams famously observed, in Western culture, "community" is a freighted term that "seems never to be used unfavorably, and never to be given any positive opposing or distinguishing term" (Williams 1983:76). Certainly, this term was one of the reasons for my initial interest in the community of Ixtlán, which was already an internationally famous example of a successful and well-organized forest community (Bray 1991; Bray et al. 2003). What were the political practices and institutions that allowed this town located in the Sierra Juárez to so precisely align itself with the desires and fantasies of the World Bank, with forestry officials in Oaxaca, and with anthropologists such as myself? This was partly due to the combination of the freighted key words "community," "indigenous," and "forests," which came together to make Ixtlán of special significance.

The mythological freight carried by the term "community" requires some clarification. Like all human institutions, real communities are politically divided and often far from egalitarian, places where the meaning and application of the term "community" are often bitterly contested. Anthropologists have long been aware that far from being egalitarian social spaces, the communities they visit and write about are often hierarchical, sometimes divided and contentious, and, in rural Mexico, often violent and oppressive (Nuijten 2003; Rus 1994). Images of harmonious and egalitarian communities are often invoked by environmental activists and community leaders (e.g., Martinez Luna 2004), and it may seem that in writing about internal political dissent, I undermine the precious political space of activists and community members. In the wake of the 1994 Zapatista rebellion, some Mexican intellectuals have attacked indigenous political movements in precisely this way, arguing that indigenous communities are corrupt and conflict ridden, that they abuse their women, and that they are unworthy of the special sympathy of urban audiences. In what follows, I will be describing political conflict and inequality in Ixtlán, but I do not present this is as a criticism of indigenous community politics. It is not at all surprising that communities have political conflicts, that they are not perfectly egalitarian, or that they can be oppressive. Indigenous and other communities should not bear the burden of fantasies about premodern pasts that make them the opposite of our present-day discontents, nor should they suffer from our disillusionment when fantasies of communitarianism or indigeneity are revealed to be an illusion. Rather than living in premodern harmony, people in Ixtlán and in other places in the Sierra Juárez use ideas of harmony and collective interest in order to organize their lives, build community institutions, and, crucially, to define a space of political autonomy and build legitimate knowledge about nature. If there is any lingering sense of unmasking, of revealing a scandalous secret of community conflict, I suggest that this is due to the fantasies and associations of these terms, to the weight of the "savage slot" (Dove 2006; Li 2000).

People in Ixtlán and the Sierra Juárez almost never spoke of indigeneity in public political pronouncements or in conversations with me. Following them, I too have little to say about indigeneity in this chapter. People in Ixtlán were clearly indigenous in the eyes of the state, and if pushed they would articulate a sense of being Zapotec and indigenous. They were undoubtedly of largely indigenous descent, and community institutions were the product of colonial and postindependence policies and legal categories that sought to regulate and control indigenous people. Nevertheless, in 2000–2001, people in Ixtlán spoke always of their identity as *comuneros* or *ciudadanos* and only very rarely of being Zapotec or indigenous. Paradoxically then, the flagship of indigenous community forestry in Mexico was run by people who seldom talked of themselves as being indigenous. This may make more sense if we follow Tania Li in seeing people's identification with indigeneity as the result of an articulation of local interests, practices, and identities with outside state and other institutions (Li 2000:151). For people in Ixtlán, logging forests, telling stories about the FAPATUX logging company, and contemporary experience of the federal forestry bureaucracy or of working in the community saw mill, were the most important practices and forms of collective identification. Indigeneity could yet become relevant in regional political organizing and alliances with other communities, but there were no conflicts that called for such alliances when I was in Oaxaca, and indigeneity was then more a spectacle than a form of political identification. This was made clear to me on the bitterly cold night of the annual *fiesta* in December 2000, when the town paid a troupe of professional dancers to perform the traditional indigenous dances of the Sierra Juárez before an audience that was at that moment more interested in watching than in performing their indigeneity.

Politics and *Cargos*

It is indicative of its freighted significance that there were two possible "communities" of which the town of Ixtlán was the center. These were the *municipio* (municipality)[5] and the *comisariado de bienes comunales* (community properties commission). Both of these had an associated public who voted for office holders and were themselves expected to hold office; this is known as holding a *cargo,* a term that means both burden and honor. Men were expected to hold successively more responsible roles in the *cargo* system, ultimately becoming elders when they had held a range of offices from most junior to most senior. This *cargo* system of community government dates back at least to colonial times and formerly combined civil and religious offices.[6] In Ixtlán, as elsewhere in Oaxaca, religious and civil office holding systems were separated from each other in the 1930s with the creation of the new legal institution of the *comunidad agraria*, with its associated *comisariado de bienes comunales* (community property commission). In Ixtlán, the *comunidad* is the legal owner of forest land and is controlled by legally registered male *comuneros* (commoners) through the *comisariado* and a set of associated office holders. The separate *municipio* administers the town and (in theory) nearby towns also, and *municipio* officials are elected by the *ciudadanos* (citizens) of the town and other neighboring dependent communities. Finally, there are religious *cargo* hierarchies attached to each of the four churches in town, as well as numerous neighborhood and school committees. Although the principal secular *cargo* hierarchies were closed to women, the various neighborhood committees actively solicited the involvement of women and young people, and most adult citizens ended up working at least on lesser committees of one kind or another.

In Ixtlán as in other parts of Mexico, *cargo* offices are the location of struggles over the control of communal institutions and the implementation of community development projects. The political structures of community government are at the interface between the state and community; leaders are responsible for negotiating with state institutions, and at different times they may decide to assert community autonomy or to ally themselves with one or another state institution. Ixtlán, for example, has long been known as an ally first of the logging company FAPATUX and more recently of the federal forest service (now SEMARNAT). Community leaders therefore face outward toward the national state and its institutions and inward toward the community, requiring them to be skilled in the art of brokerage, of making personal connections with officials in Oaxaca, of knowing how to do paperwork and activate *tramites* (bureaucratic procedures), and to be a good public speaker in community assemblies. As we shall see, not everyone managed this balancing act successfully. A critical difference between the *cargo* system and the national political institutions of the Mexican state is that participants in the *cargo* system are assessed not as citizens but as members of the community. This communal public sphere is semi-autonomous from national politics, outsiders are forbidden to take on *cargos*, and meetings may be closed to outsiders. In Oaxaca, state and federal governments do not intervene in community decisions unless there is a serious conflict, a recognition of government by *usos y costumbres* (usages and customs).[7] *Comisariados* used to be able to imprison and fine community members, but these powers have been successively whittled away by the federal government, and in Ixtlán the legal limit to community authority has diminished to putting drunks in the town jail overnight and making them clean the streets for a few hours the next morning.

Office Holding

The two principle *cargo* hierarchies had to fill a total of more than 80 offices from a pool of perhaps 400 adult men (see tables 8.1 and 8.2); this meant that at any given moment, perhaps one in five adult men occupied an office, after which they were excused duty for several years. Not all of these were full-time positions, but there were numerous addi-

Table 8.1
Municipal Government (Ayuntamiento) and Associated Police Offices in Ixtlán

Name of Office	Responsibilities	Number of Office Holders
Town Council (*Cabildo*)		
President (*presidente municipal*)	Leader of municipality	Incumbent + one reserve (*suplente*)
Treasurer (*tesorero*)	Administers municipal finances	One
Municipal secretary (*secretario municipal*)	Additional administrative work	One
Sindico		Incumbent + reserve
Regidor		Six incumbents + six reserves
Police		
Alcalde	Takes care of keys of municipal prison	Four, one is the senior
Chiefs of police (*jefes de policia*) also known as mayors of police (*mayores de policia, alcaldes de policia*)	Organizing work of the *topiles*	four, one is considered the chief of police and has a reserve or *suplente*
Policeman (*topil*)	Street cleaning, public order (may put drunks in municipal jail for the night), fetching and carrying for the municipal officials	Sixteen, four on duty at any time

tional neighborhood and school committees and for those who wished, the separate religious *cargos* associated with the church. The overall effect was to create a strong demand for young men to carry out the lower level positions such as *topil* or *alcalde de policia*, jobs in which they would keep streets clean and lock up drunks. Ideally these young men were supposed to occupy several junior positions, gradually gaining experience and ultimately ascending to the senior position of *presidente municipal* or *tesorero*. In many communities in the Sierra Juárez, heavy migration to the United States has meant that there are few working age men, creating heavy pressure on young men to occupy senior positions. In other communities, protestant evangelicals have refused to participate in *cargos*, which they associate with *fiestas* and public drunkenness. Neither of these things has happened in Ixtlán, where the *comisariado*

Table 8.2
Comunidad Offices in Ixtlán

Name of Office	Responsibilities	Number of Office Holders
Comisariado de Bienes Comunales (Commission for Community Properties) Comunidad Agraria of Ixtlán		
President of communal properties (*presidente de Bienes Comunales*)	Leader of *comunidad*	Incumbent + one reserve (*suplente*)
Treasurer (*tesorero*)	Administers communal finances	Incumbent + one reserve (*suplente*)
Secretary (*secretario de bienes comunalesl*)	Additional administrative work	Incumbent + one reserve (*suplente*)
Council for vigilance (*supervision*)		
President + two auxiliary secretariess	Supervises functioning of the *Comisariado*, audits accounts, makes sure community resolutions are implemented	Three
Advisory Commission (*Comisión Asesora*)		
Elders and professionals from forestry business	Analyzes and prepares new policies for the general assembly of *comuneros*	Twenty
Forest guards (*guardamontes*)	Responsible for patrolling forests, keeping out outsiders, detecting fires, and helping fight them	Twelve (three on duty at any one time)
Community member (*comunero*)	Elect community officials, have the right to work as loggers, cut firewood, share in profits from community forestry business	384 in 2001

is a largely secular institution, while the economic success of the logging business and the many professional services provided in town provide some employment for young men. In Ixtlán, therefore, the problem is not so much a lack of recruits as a tension between senior men's desire to recruit young men to do the lower level *cargos* and their wish to avoid having to share profits too widely.[8]

Town residents who are only *ciudadanos* (citizens) may vote in municipal elections and enter this separate *cargo* system, but those who are not also *comuneros* have no real prospect of ascending to senior positions. This meant that in a town of 2,000 people, political and financial power remained in the hands of around 300 men who controlled the profits of the forestry business and only reluctantly signed up new *comuneros*. In contrast, the *municipio* had very little money and received only modest funding from the federal government, and anyone could join in. All of these offices were substantially closed to women as only male *comuneros* were allowed to vote, and very few women came to meetings. *Municipio* meetings were open to all, but in practice only a few professional women attended. Far from being a community of equals then, Ixtlán was divided by political status and gender. Public spaces were dominated by *comuneros* who were reluctant to share power or enlist new members. Nevertheless, the growth of forest industry was changing gender roles, as women professionals began to work as accountants in the UCFAS and single mothers entered the workforce at the saw mill.

A *cargo* was both a means to rise in rank and an onerous burden with heavy financial costs. As one older woman told me:

> When I was a child the *cargos* were unpaid, my father had a full time *cargo* and in that case it was hard for the family. Everyone had to help, the family and children bore the burden, the parents, the in-laws. Now it is different because . . . cargos are paid. (Antonia Pérez, author interview notes, October 2, 2000)[9]

Even men who did not hold office were expected to attend monthly meetings that could last an entire day. In addition, all adults were expected to perform communal labor activities (*tequios*), which could range from a Sunday morning cleaning the streets to a strenuous week-long hike accompanying a government survey team along community boundaries. Older *comuneros* told me with pride that they had fulfilled their *cargos* (*cumplido con cargos*) and that this gave them the right to act in community affairs, to ask for land to build a house, and to share in the income from the forestry business.

People in Ixtlán saw the *cargo* system as something truly *of* the community; someone who had fulfilled *cargos* was respected, whereas someone who had refused a *cargo* faced disapproval and possible sanctions by the community assembly. Although most people insisted that assemblies were characterized by harmony and good feeling, even the most cursory discussion of community harmony inevitably recounted divisions.[10]

> We discuss problems in the assembly, if there is a problem, we put it to the consideration of the general assembly of *comuneros*, calm is maintained and if it is a big problem, we leave it there, we don't discuss it any more. (Ricardo Sánchez, author interview notes, September 22, 2000)

Another *comunero* emphasized that Ixtlán was unlike tempestuous and conflicted neighboring towns:

but in themselves, the people of Ixtlán are like that [calm]. For example, if we compare with Atepec, it is pure egoism there, a lot of personal attacks, but here people are more reasonable. On the contrary, when we [the office holders] don't work together people get annoyed. For example if the assembly finds out that the municipal president and the comisariado [aren't working well together] . . . then they will pull our ears [i.e. reprimand us]. . . . And at the end of the assembly everyone is like friends even if they have been rivals. (Graciano Torres, author interview notes, August 9, 2000)

Ideas about community harmony could be deployed in political debates and to accuse senior *comuneros* and professionals of "not fulfilling their *cargo*," of "taking care of their own interests," or of "only speaking to their friends." Because national parties are explicitly excluded from community politics, community political factions were instead associated with family and economic interests. I was told that the assembly of *comuneros* picked representatives of different factions to share office so that they would monitor each other. This executive of contending parties is of course quite alien to the Anglo-American political model of a unified executive and a loyal opposition; I also saw this model at work on several occasions in regional forest councils, where representatives of traditionally rival communities were voted into office. Rather than being members of a public that monitored official performance in a shared public sphere, as in models of civil society and state/society boundaries that are influential in the social sciences, people in Ixtlán trusted officeholders because they were monitored by technically skilled senior *cargo* holders who represented opposing political factions and would arrive at technical and financial decisions in relative secrecy. Knowledge about forests was credible because of the character of the senior office holders, because of their technical skills, and because of the factions they represented.

A successful senior office holder needed to have both technical and rhetorical skills, as one *comunero* told me:

In order to be president you have to have experience, to say well considered things, but you also have to be able to speak well. And even then, the president is under the advice of the *Comite de Asesores* [Committee of Advisors] who will rein him in if he goes too far . . . so not everyone gets to be president, not everyone has all of these abilities. (Juan Gomez, author interview notes, October 5, 2000)

It was difficult to balance being a professional who told forest workers what to do and a *cargo* holder who tactfully built alliances among forests, community members, and outside institutions. On occasion, this could result in spectacular failures, such as when the general assembly met in an extraordinary session to criticize the performance of the then *presidente de bienes comunales*, Antonio García. He was accused of failing to present accounts on time and of working in the interests of patrons in SEMARNAP. This was a perennial potential accusation against community and municipal presidents who were supposed to deliver alliances with outside institutions but could always be accused of being beholden to these same allies. García had been a major figure in building community forest management in Ixtlán, but he lacked affiliation with any of the dominant family/political factions in town, and in the end he felt forced to resign and moved out of town. This was something of a personal tragedy. He had dedicated his professional career as a forester to working for the community of Ixtlán. Perhaps as a result of these recent events, the new *presidente* was rather nervous about his new position and did not allow me to attend community assemblies.

Senior *cargo* holders had to reconcile responsibility toward the community with the need to build alliances with outside institutions, and they had to spread the profits from the forestry business relatively evenly, even as professionals demanded higher salaries that aroused criticism from ordinary *comuneros*. In conversations about office holders, people repeatedly mentioned the "inside" and "outside" that defined the boundary of the community, assessing the quality of leaders' alliances with outside institutions and their ability to manage finances and present regular reports inside the community; the public civic performance of community foresters was particularly important. For educated *comuneros* who were also building careers as government functionaries or professionals, it was difficult to hold down full-time civic positions. Some struggled to hold down multiple jobs or paid others to hold junior positions in their name, lying low and hoping to avoid criticism for not being physically present in their community office; others participated in community politics through powerful advisory positions rather than through holding full-time senior *cargos*.

The long history of working with the logging company meant that technical knowledge about forests was widely spread among the contending political factions. As one former municipal president told me:

> Here, if there is a technician who is from the community people trust them more, because they know that he won't be damaging us. And we have people who are experienced in forest management, they don't let themselves be fooled . . . for example with the management plan that we have now [people say] "Why does

it say this much volume [when there is less timber], the study was wrong, you are cutting too much volume here, you are affecting the land where you cut too much, it looks very strange, the slope is too steep." And that's how they question him. . . . If he tries to give a wrong technical justification so as to evade responsibility, then they catch him out "No, that's not right, you say there was a seed tree treatment, you say there should be a thousand trees per hectare." (Graciano Torres, author interview notes, August 9, 2000)

Many people in Ixtlán knew how to read the community forest management plan and how to worry about its accuracy, about the quantity of timber cut, or that logging too close to streams and springs would cause them to dry up. Thanks to a policy of paying part of the educational costs of the sons of *comuneros*, there were a number of foresters, agronomists, or other professionals who could criticize the technical quality of forest management decisions, and all of the major political factions contained such professionals. Most technical decisions were hammered out by an advisory commission (*comisión asesora*) composed of a mixture of about twenty professionals and senior *comuneros*, balancing the tension between technical knowledge and political authority based on office holding.

Building Bureaucracy in the Forest

Pine trees in the forests of Ixtlán were logged by *comuneros* who collaborated with a material document, a formidable multivolume management plan that mapped the community and its forests in exhaustive detail (TIASA 1993). This document showed rivers, streams, endangered species, ecological zones, and the all-important stand maps that located commercially valuable pine timber. Logging forests legally was a complex bureaucratic procedure that required great skill and coordination between the different community members who mapped the forest, marked trees, documented logging, prepared transport documents, and elaborate bimonthly and biannual cutting reports, as well as managing internal accounts and permits. These material practices of filling in forms, of marking, cutting, and transporting trees, stabilized and made credible the authority of the community forestry bureaucracy and gave it power in the eyes of *comuneros* and the state. The management plan was itself a material/ideological boundary object (Star and Griesemer 1989), allowing an alliance between SEMARNAP officials who saw it as a means of gaining bureaucratic knowledge and control, and community members for whom it marked success in gaining control of community forests.

Foresters had to translate scientific theories to the forests of Ixtlán; it was precisely the resistance of forests to traveling theories that gave them space to exercise their technical knowledge, to apply context-dependent local knowledge that also reached for the authority of science. Scientific theories are very poor at specifying how technologies are produced, as traveling scientific theories are transformed by local practices, alliances, and interests (Pinch and Bijker 2002). Further, the resistance of nature to human intentions becomes a material and symbolic resource, opening up a gap between scientific theories and local practices of logging and knowing forests. Foresters had to classify and map partially known forests and project their growth in space and time; they had to weave stable and credible alliances among management plans, trees, and audiences. This was at the same time profoundly extra local and profoundly context-dependent local knowledge, as foresters' knowledge linked forests in Ixtlán to the forest service and the charisma of global forestry science. Knowledge-making was not confined to foresters: Loggers and community members also translated scientific theories and used them in their practices of making sense of forests and of calling into question the knowledge of others. It was this knowledge in translation that gave people the power to retain some distance from technical knowledge and the authority of the state, and to criticize and hold foresters to account.

In Ixtlán, it was community foresters and elders who commissioned and stabilized the all-important forest management plan that legalized community control of logging, saw-milling, and furniture making, and which allowed the creation of a forest industry in Ixtlán in the late 1980s. These leaders had to enact a network that entangled trees, roads, inventory forms, and documents in order to produce the management plan that conferred the all-important right to log forests and transport timber. The title page of this plan bore the signatures of government officials and community leaders, signifying a successful political and bureaucratic alliance between state and community.

The person who told me most about this process was Antonio García, a *comunero* forester who helped broker an alliance between the community and the forest service in the late 1980s. He told me how he and the *comisariado* had commissioned a group of private foresters from Texcoco, near Mexico City, how they had to persuaded the community to pay for the plan,[11] and how they had trained teams of *comunero* assistants to walk through the forests and carry out the careful and tedious work of tallying trees. The information from tally sheets had to be analyzed by computer and collated into voluminous volumes of graphs, tables, and maps that depicted the location of valuable timber, together with narrative descriptions of soils, geology, population centers, and so on. The whole procedure called for the mastery of an enormous array of bureaucratic, technical, and political tasks by community foresters who had to build technical knowledge that was credible and legitimate to *comunero* loggers, forestry technicians, and government officials. Forest stands in Ixtlán were places that were made by the plan and by the practices of loggers and technicians, these places were linked to state and national scales by the management plan and by permissions granted by the state and federal government, and, ultimately, knowledge about these forests could be expressed by the federal government in national-level figures about timber production or forest health. This was skilled alliance building that required foresters to exercise local knowledge of what to record and not to record, of how to fill in forms and produce documents; forms and forests alike were material resources for making politics.

Foresters' principal method of tending[12] forests is to decide when, where, and how many trees to cut; they seek to encourage the regeneration of desired species and to balance ecological impacts with the production of relatively stable flows of timber, water, and other goods over the long term.[13] To help make sense of the complexity of nature, foresters map out the forests into *stands*, areas that are relatively ecologically homogenous, have similar ages and structures, and can be subjected to the same kind of logging operation. Stands are not simply found in the forest; rather they are ideal/material processes that travel through time, changing as trees grow and respond unpredictably to the intentions of foresters and loggers.[14] In practical terms, stands are usually bounded by streams, ridgelines, and other convenient landmarks. Over time, stands acquire an infrastructure of larger and smaller logging roads (see figure 8.4), while the physical and ecological impact of past logging operations can make stands easier to see, remaking the forest to suit foresters' goals of producing a legible nature.[15]

Like other classifications then, stands are conceptual categories that are both made and found in the world, as forestry technicians translate scientific theories to forests and struggle to balance the interests of loggers and timber buyers, pressures from superiors, and resistance from unruly tree species. In Ixtlán, the ecology of tree species and the quality of their timber have made some kinds of trees and forests little known. Eight thousand hectares of homogenous, widely dispersed, and easily processed pine species were carefully inventoried and logged, whereas

Figure 8.4
Working in the forest: logging trucks and loggers at the side of the road.

oak species in the same forests and stands were barely distinguished and largely ignored. Nearby, 9,000 hectares of tropical and cloud forest with much greater species diversity were completely untouched and remained uninventoried, unlogged, and little visited (see figure 8.5). This concentration on pine forest reflected the historic interests of state and commercial logging interests, but economic pressures could be deflected by community concerns and scientific knowledge. In 2000–2001, more than 723 hectares of highly productive pine forest were protected from logging because many people feared that cutting in this area could cause the failure of streams and springs; this concern drew in part on the desiccation theory that had traveled to the Sierra Juárez in the 1920s and 1930s.[16] Overall, community members were much more conservative than government foresters or forestry textbooks advocated. Loggers cut less than authorized because they disliked clear cuts, because there were no markets for oak species, and because aggressive logging risked community censure and disapproval.

I spent three weeks with community forestry technicians and loggers in the forests of Ixtlán in September and October 2000, and I later carried

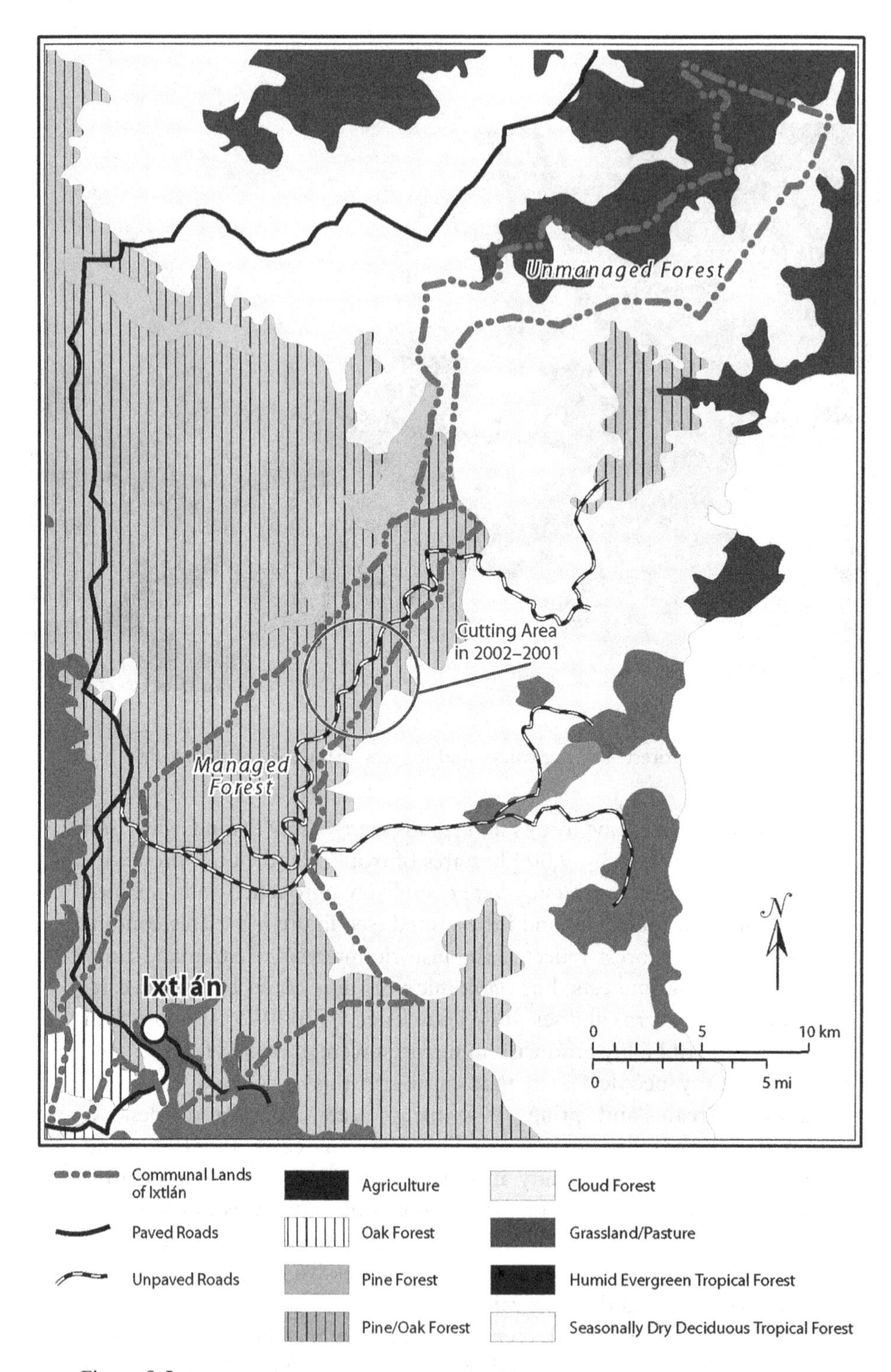

Figure 8.5
Land cover and forest management in Ixtlán. *Source:* Map by Metaglyfix.

out extended interviews with loggers and technicians at their homes in town. On a typical day, we would set out on the two-hour drive to the cutting area early in the morning. There were usually five or six of us, so we took turns sitting in the back of the pickup truck; even on sunny summer mornings, rain clouds often came up in the afternoon, drenching us all and stopping work. The journey was cold and exhausting, but we were better off than the logging crews who traveled standing up in the back of a large truck and who camped out in the forest for two or three days at a time. Technicians located stand boundaries by peering at a photocopied section of the stand map and tying this to the local landscape of roads, streams, and ridges. We had to "follow the prescription" and mark for cutting the approximate numbers and species of trees called for in the plan. We scrambled up and down steep hillsides in a loose line, marking the base of trees to be cut with blue spray paint. We constantly looked up and down and imagined future tree growth; upward at tree crowns to decide which trees to cut so as to leave an even spacing for the remainder to grow well, downward at the trunk so as to eliminate damaged or diseased trees. We also had to think about the power of senior *comunero* loggers, making sure to leave a trail of marked trees that would steer them through the whole stand and avoid protests that we had marked too little timber. Senior loggers felt free to criticize marking practices, to pressure relatively junior technicians to mark large trees and avoid small or damaged trees, and even to have an overly conservative technician banned from marking.

Once a tree was cut and stacked by the side of the road, it entered the system of official transparency and state bureaucratic knowledge; this was timber that was recorded in national forestry statistics and known to officials. At every stage, loggers, technicians, and community officials had to collaborate in filling in forms, requiring the practical, context-dependent knowledge of paper workers who negotiated what to fill in and what to leave out. Following Brenneis, it is worth tracing the careers of these forms (Brenneis 2006) in order to see how they became part of the knowledge of forests coproduced by Ixtlán and the forest service. A pile of logs at the road side had to be tallied and measured by a *recibidor*, these logs were then registered on a cutting form signed by the logging boss and a *documentador* who manned a roadside checkpoint and kept a running tally of timber trucks and their loads. Transport documents certified that logs were legally cut, allowing them to travel across roads and providing some defense from solicitations for bribes by police or army checkpoints. In Ixtlán, most timber traveled to the saw mill in town where it was transformed into planks, furniture, and pallets; these too were then elaborately documented and accompanied by forms as they traveled to market. It is the ability of Ixtlán to provide numerous workers skilled in the art of filling in forms that has made these steps possible, and it is the legitimacy of these "paper workers" in the eyes of loggers and community members that allows them to work relatively unchallenged. The skilled practice of paperwork required *documentado-res* to apply customary rules about how a form should be filled in if it was to be acceptable to officials, balancing this with pragmatic knowl-edge of logs, trucks, and political pressure from loggers and community members. All of the loggers I talked to told me of the abuses carried out by these "paperworkers" during the FAPATUX era, when collusion between a *documentador* and a truck driver could result in illicit private sales or in a logger being cheated. It was a testament to the legitimacy of community paper workers that I never heard them accused of such abuses.

One older logger, Rosendo Santos, emphasized his respect for a com-munity forest management plan that he felt was truly *of* the community, comparing this with the way that he and others had cheated to fell extra trees under the logging company, and discussed how he and his fellow loggers balanced ecological impacts and economic necessities:

> [We trust tree marking because] if there is an area we are going to exploit, we have to see if timber is coming [out of the area]. And we tell the manager if there are problems, that only thin trees are being marked, well then that is not alright. But we can't ask them to just mark big trees. . . . When we go back to look at the results where we were cutting it is all closed up again; we ourselves take care of the forest, and that's why we don't give it a clear cut, we go by the marking, by the norms of the community and by the technicians. (Rosendo Santos, author interview notes, December 8, 2000)

Loggers' technical and political skills allowed them to negotiate what went on forms, how technicians should work, and where logging should be forbidden; it was this relationship of mutual accountability that allowed knowledge on forms to be coproduced and enter official accounts, rather than being evaded or disbelieved, as it had been under FAPATUX and as takes place in many other parts of Mexico.

Alliance Building and Brokerage

In talking about Ixtlán's forests, community members constantly tried to link their forests to other places, scales, and institutions, as part of a politics of knowledge-making and career building. The dominant forestry faction has linked stands of timber in Ixtlán to the forest service in Oaxaca and Mexico City, and through World Bank reports, to international interest in community forestry and forest protection. As numerous scholars have pointed out, local places and scales are produced in relation to other places and scales (Jasanoff 2004:54–58; Tsing 2000), such scales affect the forms of political engagement in which people engage. It was the existence of these multiple scales that allowed Ixtlán's leaders to weave political alliances between forest communities, forests, and distant institutions. Other people sought to link community forests to other scales, institutions, and forms of knowledge. We have already seen the case of Pozo, who tried to undermine the professional knowledge and reforestation projects of community loggers by appealing to Nazi era forestry. This was no more than an incipient critique, but a much more powerful effort to build external alliances was mounted by a group of community professionals who were worried about the impact of logging on community water supplies. In 2005–2007, a new forester strongly pressed the community to cut diseased trees in an area of watershed protection forest. Many people feared that this could threaten community water supplies, and a loose alliance of environmentalist and biodiversity protection advocates opposed the proposed logging, calling instead for biodiversity and watershed protection. Members of this group have called for ecotourism, biodiversity protection, and alliances with national and international biodiversity protection institutions. This environmentalist faction could not deliver the economic benefits of the logging business and were unable to halt the sanitation logging, but they did connect popular understandings about forests and water supplies with the charisma and prestige of their professional knowledge.

State and Community Alliances

Ixtlán and SEMARNAP were tightly allied; in some eyes, too tightly. As a model forest community that had attracted national and international attention and prizes (Ramos 2000), Ixtlán was a shining success story that SEMARNAP officials could use to build representations of success and to seek further flows of World Bank funding. This alliance had various components, each of which required a material and ideological alliance between officials and community leaders; officials could sign management plans and transport documents and lend state approval to the quality of logging operations. Community leaders could bring *comuneros* to meetings, authorize logging in their own forests, and visibly represent the symbolically powerful community forestry sector in the eyes of international donors.

Along with other forest communities officially classified as "organized," leaders from Ixtlán were leading actors in regional forest councils and were highly visible at the forestry convention in Oaxaca in 2000. The municipal president of Ixtlán was one of the very few community members to appear on the podium, when he gave a speech on the benefits of producing an *estatuto communal* (community statute) that formalized community rules and regulations. Ixtlán was a useful model for federal officials who wished to encourage communities to become "better organized" and to make this organization more legible to the state. The president of *bienes comunales* was similarly helpful, giving a presentation about the community saw mill, while other *comuneros* had set up impressive posters outside. As we have seen, this alliance between officials and their allies in the forest communities helped stabilize representations of community industrial forestry before the hostile state governor; going along with this alliance also secured important benefits for Ixtlán's foresters, who knew they were highly unlikely to face onerous audits. Ixtlán and the neighboring forest communities could also apply pressure according to the traditional repertoire of Mexican politics, by presenting officials or politicians with a visible number of bodies, people who were prepared to block roads and occupy offices. On occasion, the communities would unite in a politically efficacious protest to press for common goals, as before the 1994 presidential elections when they threatened to blockade the highway into the city of Oaxaca and succeeded in securing an informal tax exemption that lasted at least until 2001 (ASETECO and COCOEFO 2001). Similarly, I was told that during a period of violent unrest in the city of Oaxaca in 2006, the *serranos* had told contending parties that they would remain neutral as long as their timber trucks were allowed to travel freely. Other people told me that the governor had paid large sums of money so as to secure *serranos*' passive support.

Official knowledge of timber cut in the forests of Ixtlán was therefore coproduced in a shifting alliance between officials and community leaders who could both threaten and help each other, who could collaborate in filling in and signing forms and in submitting and accepting reports and documents. Officials needed communities that could come to meetings and they feared public protests and road blockades that could cost them their jobs. Ixtlán and other forest communities needed officials who would sign timber transport documents, certify management plans, and help protect them from attempts by environmentalists to shut down logging. The volumes of timber cut in the forests of Ixtlán were recorded in official statistics not because powerful state institutions had imposed an official optic of transparent knowledge, but because powerful forest communities and relatively weak officials were able and willing to ally with each other. This knowledge was supported by a fragile network of objects and actors, of forest stands and management plans, of trees and loggers in the forests, of timber transport forms, and of the saw timber they accompanied to market. Illegally logged and sawed timber lacked official signatures and accompanying documentation and did not enter official statistics, the large-scale national categories of timber production and of official order and knowledge.

It would be easy to see the alliance of state and community as the imposition of an oppressive bureaucratic project of official legibility (Scott 1998) or perhaps as the *comuneros*' internalization of an official project of environmental governmentality (Agrawal 2005). Certainly, people in Ixtlán were consummate bureaucrats, with regard to both the state and their own internal permissions and accounts. However, theories of governmentality or authoritarian bureaucracies gone wild obscure the skillful and context-dependent knowledge of community paperworkers and loggers who filled in forms and logged forests, and who translated bureaucratic classifications and foresters' theories into local practices. Official forms of knowledge and work are never applied without transformation and translation in local contexts of use; scientific theories are transformed in the making of technologies. This gap between a traveling scientific theory or bureaucratic regulation, and the contexts where it was actually used, allowed loggers to remake themselves as political and epistemic actors who could criticize official knowledge or frustrate efforts to count trees and map forests.

Community members could also use their epistemic and political power to make themselves illegible to other state institutions. In 2000–2001, the federal government was rushing to complete a cadastral survey of all property boundaries in Mexico, in an ambitious program called PROCEDE. Many rural communities refused to participate, either because they had outstanding disputes with their neighbors or because they did not wish the government to know who owned land within community boundaries. People in Ixtlán clearly saw the PROCEDE program as useful for stabilizing communal land tenure: The surveyors were accompanied around the boundaries of community lands, but they were not allowed to map private property within these boundaries. The fractured nature of the state allowed community leaders to form alliances with some institutions (SEMARNAP), to repel others (PROCEDE), and to produce official ignorance about who owned property in town. This resulted in the paradoxical situation that detailed official knowledge (for SEMARNAP) existed regarding the boundaries of distant stands within the forests of Ixtlán while PROCEDE was totally ignorant as to who owned house plots in the town center.

Conclusions

In this chapter, I have argued against the idea of community as a space of unity and assent and tried to give something of the flavor of community politics and of how community leaders build careers by continually seeking to build alliances between local people, local places, and outside institutions and scales. Ixtlán is an unequal place: Men have more power than women, *comuneros* have more access to land and money than ordinary *ciudadanos*, and professionals have more chance of ascending the *cargo* system than do the less educated. Nevertheless, in Ixtlán, the community is also a public space where the technical knowledge of forestry professionals can be questioned. Professionals can be held accountable not only for their technical competence but as office holders with political obligations toward the community. Their role as intermediaries between community and state makes these professionals vulnerable to criticism and sensitive to concerns about how to log the forest and distribute profits. It is this sense of vulnerability that makes professionals accede to community members' criticisms; these experts' technical knowledge about community forests is credible to community members because those who produce it can be called to account. This accountability is not simply a demand for a professional to produce information in public; rather it can result in a professional losing his job. Rather than being unified and harmonious then, Ixtlán is a place where ideas about harmony and community are used to make professionals in some measure technically and morally accountable.

Technical knowledge of forests is also made credible within the community because it is certified and monitored by contending political factions that have technically trained representatives of their own.

Community trust in forest management is produced by relatively equal contending factions, rather than being imposed by one group alone. In many communities across Mexico, this is impossible; technical knowledge about forests and paperwork is too scarce, and a single faction or person can monopolize knowledge, act as political brokers with the outside world, and use forests for their own benefit. Such communities are often plagued by violent conflicts and illegal logging (Klooster 2000).

Following scholars of Science and Technology Studies, I have described knowledge-making as a shifting set of social practices that continually seek to entangle people and things.[17] These difficult practices require creative work; they are also practices that encounter political difference and hierarchy, where knowledge is made by people who can call each other to account. This begins to make it clear why Ixtlán, unlike most forest communities in Mexico, was able to log its forests legally, why it was able to collaborate with SEMARNAP officials, and why the Mexican government knew about logging in the remote forests of the Sierra Juárez. Within the political context of state–community relations in Mexico, it was the political, material, and representational power of forest communities such as Ixtlán that allowed for an entanglement of the social worlds of World Bank forestry projects, national forest policies, and Ixtlán's decision to produce management plans, submit logging records, and attend the forestry convention in Oaxaca. People from the vast majority of relatively weak or less unified forest communities, where bureaucratic practices of forest management lack popular assent, are likely to adopt more traditional practices of evasion, concealment, or illegal logging.

People in Ixtlán continually thought about their relationship to other places; this was part of the practical politics of contesting the authority of official knowledge, of reaching alternative allies and institutions. They asserted local autonomy even as they linked local scales and forms of nature to external institutions and scales. Jose Pozo's discussion of Rodolfo Hitler in the name of advocating burning and against reforestation was therefore analogous to discussions of global climate change and biodiversity protection by environmentalists who advocated protected areas and alliances with biodiversity protection institutions. Political entrepreneurs in Ixtlán who wished to build successful careers had to build material and conceptual alliances that linked local scales and forms of nature with other places and scales, using the charisma of global environmental change or Nazi forestry and linking it with people, documents, and forests. Rather than flowing seamlessly from the state, to the management plan, to the beliefs and identities of community members, knowledge was made and negotiated, it became a political resource precisely because it was coproduced in encounters between loggers and forests, between foresters and *comuneros*, and in negotiations between community leaders and officials. The precise effects of a traveling discourse such as forestry science depends therefore on the entanglement between institutions, discourses, local politics, and skillful practices of logging in forests. Public assent to industrial forestry in Ixtlán was produced not by the imposition of a state project of legibility and of official forms of knowledge, but by the ability of a well-organized and powerful community to collaborate in making knowledge that would inhabit national timber production statistics and official reports. Work in the forests as loggers and forestry technicians, marking community boundaries and burning diseased trees, has provided community members with a detailed technical knowledge of forests that they can use to criticize the authority of community and state foresters. Crucially, this is knowledge in translation; theories derived from forestry science encounter the material resistance of real nature, they are applied to forests by evasive and stubborn people; it is this translation and dislocation that makes knowledge about forests a political resource for community members who wish to undermine or engage with official projects of knowledge-making.

State forestry science and policy have not encountered passive local subjects in the Sierra Juárez. On the contrary, knowledge of forests is woven—it is an entanglement among traveling theories, fractured states and communities, and continually shifting alliances of humans and nonhumans. As Sheila Jasanoff points out, socially accepted knowledge is the product of practices of participation, deliberation, and representation; her coinage "civic epistemologies" (Jasanoff 2005) draws attention to the importance of public assessments of claims to scientific knowledge and expertise, and it underlines the importance of public participation in knowledge-making. In Ixtlán, it is the articulation between community politics and state knowledge-making that has produced knowledge about forests that is credible and legitimate to community members. It would be easy to think that forestry is a global science, a discourse that remains in the hands of distant officials who seek to impose it on more or less passive local actors. The travels of forestry science into the forests of Ixtlán show on the contrary how people can domesticate science and turn it against the state. Further, we see here how what appeared to be small-scale, far away, and insignificant details turned out to affect the stability of such large things as the state and the credibility of scientific knowledge about forests. Other sciences and other projects of traveling knowledge may be similarly remade and undermined by unruly people and recalcitrant natural actors; all projects of transparency turn out to be profoundly mediated by people, institutions, and places.

9

Conclusion

This book has described how the science of forestry traveled to Mexico, how it came to be institutionalized by the expanding Mexican forest service, and how, ultimately, forestry was domesticated and turned against the state by particular indigenous forest communities in the state of Oaxaca. By traveling the length and breadth of Mexican forestry in time and space, I have found something that was surprising to me. Far from being an expanding and authoritative bureaucracy, the Mexican forest service, in its various guises, was hesitant, faltering, and frequently reconfigured. Forestry officials had to deal with competing institutions, skeptical publics, and unruly indigenous communities. The physical limitations of where officials could locate offices and whether they could arrive in distant communities or man checkpoints all made a difference to the building of the forest service and its forms of knowledge. This suggests something important about how we might look at other knowledge regimes in other fields entirely: We need breadth in time and space if we are to be able to notice the kinds of hesitations, reversals, and institutional reconfigurations that I have shown were a feature of the Mexican forest service. Here, history might reveal that the most authoritative institutions of modernity are less solid, enduring, and rigid than we assume, and that when we see states as performances that are continually in the making, we find them to be livelier and more surprising than we had thought them to be. Certainly, this is a bold conjecture: We need other historically informed investigations of knowledge institutions if we are to find out whether, in other places and at other times, the appearance of authoritative and knowledgeable institutions reflects the collective imagination of desirable futures rather than lived realities.

In Mexico, the knowledge regime of forestry expanded not only because of powerful institutions but because of the instability, vulnerability, and weakness that made officials, foresters, and indigenous people in some measure mutually accountable, vulnerable to each other in various ways. The project of knowing forests succeeded not because all involved were persuaded, but because they were not fully persuaded. The gap between official knowledge about forests and the kinds of things that loggers or forestry officials knew allowed for a certain critical distance on projects of public knowledge. This gap was a space for a tactics of making alternative knowledges and of opposing or remaking forestry institutions. This finding, it seems to me, applies to ambitious projects of knowledge more widely, from state projects of environmental control to projects that aim to inventory world carbon stocks, and reduce green house gas emissions through the trading of carbon offsets in international carbon markets. All of these are projects that seek to perform transparent knowledge of nature, and all will encounter practices of obscurity and concealment, the political contexts where officials decide what to enforce, and on whom, the moments where peasants or loggers decide what to write down on a document and what to leave off. Here too, I suggest, stable forms of public knowledge and stable environmental institutions will be underpinned not by uniformly shared public knowledge and powerful institutions but by the coexistence of multiple forms of knowledge and nonknowledge, and by the mutual vulnerabilities of institutions and their audiences.

The uneven expansion of a particular state institution (the Mexican forest service) and of a particular type of scientific knowledge (forestry and conservation) is an example of the more general finding that knowledge is always made alongside various forms of nonknowledge. Official knowledge of forests inescapably suppresses and silences other forms of knowledge. The intimate doubts of officials themselves, their collusion in agreeing that some kinds of things are best left unmentioned in public events, and the shared assumptions of politicians and urban audiences all come together to sustain performances of official reasonableness and of irrational or environmentally degrading rural people. This silencing, however, is not a scandal but an inescapable result of the ways that knowledge is built and sustained. Silencing and suppressing alternative forms of knowledge is not pathological, nor a sign of deficit, of the particular incompetence or corruption of Mexican forestry officials, nor is it the result only of a relatively weak and poorly funded forestry bureaucracy. Here, I can do no more than gesture to broader literatures that describe other bureaucracies, working at other times and in other places. In their different ways, many scholars have come to agree that bureaucracies which claim to speak for authoritative knowledge inevitably silence or suppress rival forms of knowledge, whether at the conceptual boundaries that separate institutions (Bowker and Star 1999; Douglas 1986), within institutions through practices of audit and accounting (Brenneis 2006; Mosse 2005; Power 1997; Strathern 2000), or in collusion between officials and their clients, as I have shown in this book.

Once we see knowledge and nonknowledge as going together, the importance of thinking about knowledge-making and state-making as more or less dramatic and unstable performances becomes clear. The texture of knowledge-making performances matters not only because these are fascinating moments of encounter but because the textures of encounters can come to affect the legitimacy of the state or the credibility of public knowledge. In the Mexican case, this means that officials have to pay great attention to the texture of their performances of authoritative knowledge. They have to pay attention to how the audience knows, to the audience's expectations of how state authority is to be performed, how knowledge is to be declared, and what forms of opposition and public declarations the audience might make. Although I have focused on Mexico, I suggest that this finding is applicable to other knowledge institutions, working at other places and times. All bureaucracies, however powerful, face certain constraints. Like the Mexican forest service, other bureaucracies have more mandates than resources; they too will have an audience of rival institutions, of keenly watching and hungry officials, of opportunistic politicians, of dueling experts and of skeptical citizens. These variably constituted and largely self-chosen audiences are likely to be watching for a sign of loss of grip, of a weakness that could undermine an institution or destroy a career. Precisely what constitutes a mistake and what kinds of knowledge performances are considered to be failures will of course vary greatly depending on history and context. Following Sheila Jasanoff (2005:247–271), we could call such differences "civic epistemologies," culturally specific collective framings of the proper way of making public knowledge and legitimate institutions. This term, which refers to the power of citizens as knowers, can also accommodate culturally specific understandings of nonknowledge, collusion, and ignorance, which live with and sometimes sustain public knowledge. Here the metaphor of performance makes it clear that what is not officially known may nevertheless be partially known; it also makes it clear that the audience collaborates in partial knowing and not knowing.

In this book, I have argued that at times the local contexts of knowledge-making matter a great deal, because the texture of apparently local knowledge performances turns out to affect what people think the state is and whether they trust official knowledge about forests. The reason that the Mexican state fails to record most agropastoral fires is that officials and farmers collude in avoiding or evading burning regulations. Ethnographic, empirical detail about what is said or not said in encounters between officials and audiences turns out to explain the meanings of national fire statistics and also, in some measure, to explain why many Mexicans distrust the beneficence and authority of the state that produces such statistics. Here I follow a number of scholars who point to the power of local-level bureaucrats to decide whether something does or does not fit a particular rule, whether of accounting or rule breaking (MacKenzie 2008). No rule or regulation contains sufficient information to explain all the ways in which it can be applied, and the contexts of application inevitably depend on relatively lowly officials, ordinary bureaucrats in small offices filling in mundane documents that are ultimately collated into national statistics or company financial reports. An empirical investigation of the pragmatics of form filling (what to leave in, what to leave out), and of what is said and not said, tells us much about the meaning and content of national and international statistics, whether about forests, fisheries, or other fields entirely. The current financial crisis was in part caused by the collusion of mortgage lenders and borrowers in filling out documents that ended up being fraudulent or dangerous, and this fact suggests that the textures of local knowledge and ignorance may be fruitfully explored around many other kinds of knowledge projects.

It has long been known that in certain cases international statistics about fisheries were drastically affected by the particular bureaucratic imperatives faced by fisheries officials. In late Soviet Russia, officials were compelled to report much lower numbers of whales killed than was really the case; more recently, Chinese provincial fisheries officials reported increases in fisheries in order to comply with mandates from their superiors (Pauli 2001). Such falsifications are found not only in non-Western places; we have only to think of the statistics of educational success produced by teachers who face demands to demonstrate that their students have learned, of police departments pressed to demonstrate falling crime rates, and of nation states that struggle to demonstrate falling rates of unemployment. In each of these cases, the meaning of "large things"—national statistics and the legitimacy of the state—turn out to depend on things that we typically take to be small and relatively unimportant, on the texture of meetings between officials and their clients. Anthropologists and like-minded social scientists can look at such encounters, such places of knowing and not knowing, and we can demonstrate that local context and ethnographic detail matters not only because it is fascinating and we love it, but because such details can matter for the careers of the powerful and the destinies of institutions.

Some centuries ago, a mundane domestic object became important for allowing a restive and skeptical public first to criticize and ultimately to rebel against the authority of an apparently well-established king. It was an insignificant object, literally, almost unnoticeable and unmentionable, a warming pan usually used to warm the royal bed. Such objects are not usually mentioned by historians, lawyers, and other people who follow the destinies of kings and states. Some readers will recognize the king as James II of England and the warming pan as the object that had supposedly been used to smuggle a boy baby into the royal bedroom, allowing James to claim that his Catholic wife had given birth to a male heir. This mundane domestic object was to feature in engravings, songs, debates, and arguments: A small deal had become a big deal. All bureaucracies and all knowledge projects face this conundrum of lively people and things, which can too easily escape their proper slot. Such events confirm the finding that scales are made and not found and that scales are made in relation to each other (Jasanoff 2004c; Tsing 2000). It is this potential slipperiness of scales that makes it unclear, beforehand, what is a small detail and what is a big deal. This suggests that scandals, moments of erupting doubt, and the remaking of institutions are something we should think about more. I pose this as a question rather than a conclusion, but Sheila Jasanoff's work on the BSE (bovine spongiform encephalopathy) controversy in the United Kingdom is particularly suggestive in this regard (Jasanoff 2005).

The uneven, halting, and hesitant journey of forestry science into indigenous forest communities provides clues for how we might think about other expanding knowledge projects and institutions. For example, we might try to trace the movement of the neoliberal economic theories that have driven political and economic reforms in states around the world, or we might look at the recent destiny of climate change science, which has affected states around the world but has yet to be anchored in local practices and economic transformations. Critics of neoliberalism have, it seems to me, too easily assumed that states are authoritative institutions that seamlessly transfer economic theories into detailed daily practices in markets and government offices. I reluctantly include in my disagreement the many scholars who have worked to criticize "neoliberal conservation," highlighting the burgeoning of conservation initiatives that are premised on environmental markets, increased national parks, or more green spaces on the map. As much as I sympathize with the thrust of this body of research, I feel that it may overemphasize the coherence and material reach of state institutions in general, and of conservation institutions in particular. I agree that discourses of neoliberal conservation have indeed spread out through the world, but, like the science of forestry in twentieth-century Mexico, I doubt that this is as smooth or seamless as is often assumed. On the contrary, I suggest that conservation officials must carefully calibrate their judgments as to what official discourses really mean, how environmental markets are to be produced, and how much they must defer to the understandings and concerns of their particular audiences. I wonder if, like the Mexican forestry officials I describe, conservationists who promote environmental markets may also inhabit unstable and spatially patchy bureaucracies, whether they too may be doubtful, vulnerable, and hesitant, and whether they too may choose to collude in not knowing the evasions and failures that their knowledge-making lives with.

Finally, let me turn to the question of what it means for the project of knowledge and of studying knowledge, if we believe that knowing and not knowing travel together, that each form of knowledge lives against what is partially known and what was not allowed to be known, against the lively alternative forms of knowledge that might yet emerge to displace their rivals. It seems to me that this changed understanding of the anthropology or social science of knowledge also changes our political engagements and theoretical commitments. For me these are new vistas; they force me to think in new ways, toward which I can do no more than gesture. I offer these ideas not as conclusions, then, but as questions unanswered, avenues that might be fruitful, ideas that might be useful to others.

First of all, it seems to me that once we see knowing and not knowing as linked and often as co-constitutive, we must reconsider our commitment toward knowing, finding, and revealing how knowledge is made, and we must pay much more attention to the various forms of nonknowledge that accompany the knowing practices of our subjects. In addition, we have to rethink our traditions of denouncing the suppression of alternative and valuable forms of knowledge, perhaps in the regimes of authoritarian states (Scott 1998), perhaps when development institutions ignore local history or cultures (Ferguson 1994), and even when development institutions hire anthropologists and bleach out their knowledge into official simplifications and policies (Li 2006). These kinds of criticisms describe a scandal of official knowledge that erases important local details. If we start from a different premise, that all knowing is accompanied by the suppression of alternative forms of knowledge, we have a different criticism to make and a different way of expressing it. The proper criticism of official simplifications is not that they suppress alternative knowledges (a more or less knee jerk argument with which most bureaucrats would agree) but to ask about the relationship between knowledge and nonknowledge. We need to ask who knows, with what authority, and with what moral, political, and environmental consequences, but we also need to focus on the nonknowledges that travel more or less silently alongside official knowledge or are perhaps wrapped into government documents and reports. Here we have to resist the urge to engage in a critique of ideology and to focus instead on the pragmatic details, the hesitant judgments of officials, and the forms of assent or the reversals that such judgments may encounter. Our descriptions of such pragmatic judgments must be symmetrical, we must also acknowledge the limits of our own knowing practices, and we must write the limits of knowing into our own practice of describing the world.

Thinking with performance and nonknowledge leads us also to think very differently about the role of the social sciences in studying knowledge, most especially when it is knowledge held by powerful officials or eminent scientists. Descriptions of the knowledge practices of others that explain away their knowledge as ideological are inherently asymmetrical, as has been pointed out by Bruno Latour and others (Latour 1993, 2004). If we take the knowledge performances of others to be just that, potentially unstable performances, we have become involved in these performances as critics, witnesses, or participants. We are offering possible interpretations that might affect knowledge-making by officials. If Mexican citizens and officials are persuaded by some aspect of my argument, they might change how they understand the Mexican state and how they understand official statistics about forests. If this is so, we can no longer absolve ourselves from being implicated in the projects of those we describe; our own descriptions and imaginations are potentially powerful practices that might make new things in the world, new kinds of institutions, forests, or forms of knowledge. Taking imagination seriously means abandoning the luxury of "seeing" what we describe, with its promise of showing all that there is to know, without however being involved in making and doing. Once we realize that seeing is material, involved, and implicated, our descriptions are less innocent but possibly more powerful and more generative. We can take seriously the power of imagination to make worlds that are more livable for officials, indigenous people, and trees.

Appendix

All names in this appendix are pseudonyms unless they are starred (*). Interviewees not quoted in this book have no pseudonym listed here.

Name	Interviewee	Date	Location
—	Academic	20-Aug-08	Mexico City
—	Academic	22-Feb-01	Mexico City
*Antonio Azuela	Academic	15-Jul-08	Mexico City
—	Academic	21-Jul-09	Mexico City
—	Academic	25-Jul-09	Mexico City
*Carlos Gay	Academic	7-Aug-09	Mexico City
Gomez, Santiago	Academic	16-May-00	Oaxaca
Gomez, Santiago	Academic	19-May-00	Oaxaca
López, Antonio	Academic	22-May-00	Oaxaca
—	Academic	4-May-01	Oaxaca
—	Academic	1-Aug-08	Oaxaca
—	Academic	31-Jul-09	Oaxaca
Sierra, Facundo	Academic, Ex Government	12-Apr-00	Mexico City
*Ludington, Stephen	Academic, Geologist	14-May-01	Oaxaca
—	Comunero/a	27-Jul-02	Ixtlán
—	Comunero/a	25-Jan-01	Sierra Juarez
—	Comunero, forester	9-Aug-00	Ixtlán
—	Comunero, forester	25-Jan-01	Ixtlán
Arce, Aureliano	Comunero/a	25-Jul-00	Ixtlán
Pérez, Zenaido	Comunero/a	25-Jul-00	Ixtlán
—	Comunero/a	28-Jul-00	Ixtlán
Echeverría, Miguel	Comunero/a	28-Jul-00	Ixtlán
Pozo, Jose	Comunero/a	3-Aug-00	Ixtlán
—	Comunero/a	8-Aug-00	Ixtlán

Name	Interviewee	Date	Location
Torres, Graciano	Comunero/a	9-Aug-00	Ixtlán
—	Comunero/a	9-Aug-00	Ixtlán
—	Comunero/a	11-Aug-00	Ixtlán
—	Comunero/a	11-Aug-00	Ixtlán
—	Comunero/a	12-Aug-00	Ixtlán
—	Comunero/a	12-Aug-00	Ixtlán
—	Comunero/a	14-Sep-00	Ixtlán
Pérez, Miguel, and Friends	Comunero/a	14-Sep-00	Ixtlán
Echeverría, Miguel	Comunero/a	18-Sep-00	Ixtlán
—	Comunero/a	21-Sep-00	Ixtlán
—	Comunero/a	22-Sep-00	Ixtlán
—	Comunero/a	22-Sep-00	Ixtlán
—	Comunero/a	2-Oct-00	Ixtlán
—	Comunero/a	2-Oct-00	Ixtlán
—	Comunero/a	2-Oct-00	Ixtlán
—	Comunero/a	2-Oct-00	Ixtlán
Pérez, Zenaido	Comunero/a	3-Oct-00	Ixtlán
—	Comunero/a	5-Oct-00	Ixtlán
—	Comunero/a	11-Oct-00	Ixtlán
—	Comunero/a	14-Oct-00	Ixtlán
—	Comunero/a	16-Oct-00	Ixtlán
—	Comunero/a	16-Oct-00	Ixtlán
—	Comunero/a	23-Oct-00	Ixtlán
—	Comunero/a	23-Oct-00	Ixtlán
—	Comunero/a	5-Dec-00	Ixtlán
—	Comunero/a	8-Dec-00	Ixtlán
Pérez, Bulmero	Comunero/a	8-Dec-00	Ixtlán
—	Comunero/a	8-Dec-00	Ixtlán
Ramírez García, Luis	Comunero/a	11-Dec-00	Ixtlán
—	Comunero/a	17-Jul-01	Ixtlán
Pérez, Zenaido	Comunero/a	24-Jul-02	Ixtlán
—	Comunero/a	24-Jul-02	Ixtlán
—	Comunero/a	25-Jul-02	Ixtlán
Pozo, Jose	Comunero/a	25-Jul-02	Ixtlán
Pérez, Epifanio	Comunero/a	25-Jul-02	Ixtlán
Torres, Graciano	Comunero/a	25-Jul-02	Ixtlán
Ruíz, Abelardo	Comunero/a	25-Jul-02	Ixtlán
Ruíz, Abelardo	Comunero/a	27-Jul-02	Ixtlán
Zenaido Pérez and Josefina Pérez	Comunero/a	28-Jul-02	Ixtlán
—	Comunero/a	29-Jul-07	Ixtlán
Pérez, Zenaido	Comunero/a	7-Aug-08	Ixtlán

Name	Interviewee	Date	Location
Pérez, Josefina	Comunero/a	7-Aug-08	Ixtlán
Arce, Aurelio	Comunero/a	7-Aug-08	Ixtlán
—	Comunero/a	8-Aug-08	Ixtlán
Torres, Graciano	Comunero/a	8-Aug-08	Ixtlán
—	Comunero/a	8-Aug-08	Ixtlán
Arce, Aurelio	Comunero/a	9-Aug-08	Ixtlán
Sánchez, Antonio	Comunero/a	1-Dec-00	Oaxaca
—	Forest Forum	26-Apr-00	Oaxaca
—	Forest Forum	28-May-00	Oaxaca
—	Forestry Forum	26-Sep-00	Ixtlán
—	Forestry Forum	28-Sep-00	Ixtlán
—	Forestry Forum	24-Oct-00	Ixtlán
—	Forestry Forum	8-Nov-00	Oaxaca
—	Forestry Forum	21-Nov-00	Oaxaca
Sánchez, Abraham	Forestry Service Provider	1-Jul-98	Ixtlán
Rodríguez, Eduardo	Forestry Service Provider	4-Jan-01	Oaxaca
Ruíz, Mauro	Forestry Service Provider	24-Jan-01	Oaxaca
Ruíz, Gustavo	Forestry Service Provider	14-May-01	Oaxaca
Rodríguez, Eduardo	Forestry Service Provider	15-May-01	Oaxaca
—	Forestry Service Provider	16-Jun-01	Oaxaca
Porrua, Jose	Forestry Service Provider	20-Jun-01	Oaxaca
Rodríguez, Eduardo	Forestry Service Provider	26-Jun-01	Oaxaca
—	Government, CONABIO	22-Aug-08	Mexico City
—	Government, CONABIO	7-Aug-09	Mexico City
—	Government, CONAFOR	22-Jul-09	Guadalajara
—	Government, CONAFOR	22-Jul-09	Guadalajara
Porrua, Jose	Government, CONAFOR	30-Jul-02	Oaxaca
Domínguez, Antonio	Government, SEMARNAT	30-Jul-08	Oaxaca
—	Government, CONAFOR	30-Jul-09	Oaxaca

Name	Interviewee	Date	Location
—	Government, CONAFOR	1-Aug-09	Oaxaca
Domínguez, Antonio	Government, SEMARNAT	1-Aug-09	Oaxaca
*Azuela, Antonio	Government, Ex director of PROFEPA	21-Feb-01	Mexico City
*Azuela, Antonio	Government, Ex director of PROFEPA	23-Feb-01	Mexico City
*Azuela, Antonio	Government, Ex director of PROFEPA	25-Feb-01	Mexico City
*Azuela, Antonio	Government, Ex director of PROFEPA	23-May-01	Mexico City
Mares, Jose	Government, Ex SEMARNAP	5-Apr-01	Mexico City
*Carabias Lilo, Julia	Government, Ex SEMARNAP	1-Jul-01	Mexico City
—	Government, Ex SEMARNAP	15-Nov-00	Oaxaca
Cruz, Martin	Government, Ex SEMARNAP	5-Apr-01	Mexico City
—	Government, Federal Deputy	25-Jul-08	Mexico City
*Zedillo, Ernesto	Government, former President	6-May-03	New Haven, Connecticut
—	Government, geologist	2-Sep-01	Oaxaca
—	Government, INE	16-Aug-08	DF
*Julia Martinez+ Subordinates	Government, INE	23-Jul-08	Mexico City
*Edmundo Blanco-INE	Government, INE	23-Jul-08	Mexico City
*Julia Martinez	Government, INE	26-Jul-08	Mexico City
—	Government, INE	24-Jul-09	Mexico City
—	Government, INIFAP	4-Apr-01	Mexico City
Negrete, Paco	Government, INIFAP	19-Apr-00	Oaxaca
—	Government, PROFEPA	16-Jan-01	Oaxaca

Name	Interviewee	Date	Location
Alarcon, Rosendo	Government, PROFEPA	26-Jun-01	Oaxaca
Alarcon, Rosendo	Government, PROFEPA	23-Jul-01	Oaxaca
—	Government, Registro Agrario Nacional	12-Jan-01	Oaxaca
—	Government, SEMARNAP	24-Jul-02	Ixtlán
—	Government, SEMARNAP	16-Jun-98	Mexico City
Pacheco, Gonzalo	Government, SEMARNAP	10-Apr-00	Mexico City
—	Government, SEMARNAP	11-Apr-00	Mexico City
Cruz, Martín	Government, SEMARNAP	16-Apr-00	Mexico City
Arguelles, Miguel	Government, SEMARNAP	4-Apr-01	Mexico City
—	Government, SEMARNAP	5-Apr-01	Mexico City
Domínguez, Aldo	Government, SEMARNAP	26-Apr-00	Oaxaca
—	Government, SEMARNAP	17-May-00	Oaxaca
Germán, Ipolito	Government, SEMARNAP	31-May-00	Oaxaca
Germán, Ipolito	Government, SEMARNAP	27-Nov-00	Oaxaca
Mecinas, Luis	Government, SEMARNAP	29-Nov-00	Oaxaca
—	Government, SEMARNAP	29-Nov-00	Oaxaca
González, Roberto	Government, SEMARNAP	30-Nov-00	Oaxaca
—	Government, SEMARNAP	5-Jan-01	Oaxaca
Prieto, Romualdo	Government, SEMARNAP	9-Jan-01	Oaxaca
—	Government, SEMARNAP	16-Jan-01	Oaxaca
Domínguez, Aldo	Government, SEMARNAP	4-Jul-01	Oaxaca

Name	Interviewee	Date	Location
Soriano, Miguel	Government, SEMARNAP	7-Jul-01	Oaxaca
—	Government, SEMARNAP	27-Apr-00	Veracruz
*Fernando Tudela	Government, SEMARNAT	16-Aug-08	DF
—	Government, SEMARNAT	19-Aug-08	DF
—	Government, SEMARNAT	4-Apr-01	Mexico City
—	Government, SEMARNAT	4-Apr-01	Mexico City
—	Government, SEMARNAT	25-Apr-01	Mexico City
Colón, Miguel	Government, SEMARNAT	25-May-01	Mexico City
—	Government, SEMARNAT	23-Jul-08	Mexico City
—	Government, SEMARNAT	21-Jul-09	Mexico City
Germán, Ipolito	Government, SEMARNAT	26-Apr-00	Oaxaca
—	Government, SEMARNAT	16-Jan-01	Oaxaca
—	Government, SEMARNAT, meeting	25-May-01	Mexico City
—	Government, SEDAF	4-Jan-01	Oaxaca
—	Government, SEDAF	5-Jan-01	Oaxaca
—	Government, Institute of Ecology	21-Jan-01	Oaxaca
—	Lawyer	7-Feb-01	Mexico City
—	NGO	14-Dec-00	Ixtlán
—	NGO	18-Dec-00	Ixtlán
—	NGO	6-Jul-00	Mexico City
—	NGO	21-Feb-01	Mexico City
—	NGO	5-Aug-09	Mexico City
—	NGO	6-Aug-09	Mexico City
—	NGO	6-Aug-09	Mexico City
—	NGO	7-Aug-09	Mexico City
—	NGO	8-Aug-09	Mexico City

Name	Interviewee	Date	Location
*Hernández, Ricardo	NGO, World Bank	8-Aug-09	Mexico City
—	NGO	25-May-00	Oaxaca
Rosas, Juan and Colleague	NGO	29-May-00	Oaxaca
—	NGO	29-May-00	Oaxaca
—	NGO	30-Nov-00	Oaxaca
—	NGO	12-May-01	Oaxaca
—	NGO	13-Aug-08	Oaxaca
—	NGO	30-Jul-09	Oaxaca
*Rodolfo Lacy	NGO, Centro Mario Molina	22-Aug-08	Mexico City
—	NGO, Oaxaca	8-Jan-01	Oaxaca
Rosas, Juan	NGO, Oaxaca	1-2-May-01	Yavesia

Notes

Chapter 1

1. Author transcription and translation. The screenplay records a slightly different version of this scene (Zavattini 2005).

2. This also clarifies the awkward conundrum of discursively powerful but materially weak conservation and development institutions. Earlier work on international conservation/development noted the seamlessness of conservation/development discourse. More recently, researchers have noted that conservation and development institutions face local opposition and are often quite weak.

3. For a similar argument about the power of financial markets, see Karen Ho's description of Wall Street investment bankers (Ho 2005).

4. Beginning in the early 1980s, community-based natural resource management (CBNRM) aroused enormous academic and policy interest around the world. This was the result of a conjunction of popular environmental protests against coercive state bureaucracies, with declines in state resources in the face of neoliberal reforms. Ostrom describes the theoretical basis for CBNRM (Ostrom 1991); for a comprehensive review of common property research, see Agrawal (2001).

5. Western/Northern understandings of indigeneity are often based not on real indigenous people but derived from the presumed opposition between indigenous people and modernity (Brosius 1997; Li 2000).

6. In this book all names are pseudonyms unless I clearly state to the contrary.

7. FAPATUX is the initials of *Fábricas Papeleras de Tuxtepec* (i.e., the Tuxtepec Paper Factories).

8. For a discussion of state-making in Mexico, see Joseph and Nugent (1994). For an excellent recent review of the literature on state-making, see Sharma and Gupta (2006).

9. On the idea of development as a neutral technical process, see Ferguson (1994).

10. For a classic exposition of this actor-network theory of scientific knowledge, see Latour (1987). An elegant example of this kind of analysis is found in Timothy Mitchell's discussion of malaria in Egypt (Mitchell 2002:19–53).

11. The term "epistemic community" was introduced by Peter Haas (Haas 1990) to refer to the international community of environmental scientists working on the Mediterranean in the early 1970s.

12. The institution or institutions responsible for forests has changed repeatedly over the last century. For convenience I will use the term "forest service" to describe the federal bureaucracy responsible for forests. However, at the time of my fieldwork in 1998–2001, the primary government agency was SEMARNAP, the *Secretaría de Medio Ambiente, Recursos Naturales y de Pesca* (Ministry of Environment, Natural Resources and Fisheries).

13. See, for example, the rich literature on audit cultures in the UK (Power 1997; Strathern 2000).

14. I owe this insight to the art historian Alex Nemeroff.

15. The idea of public illusions that conceal the dangerous reality of politics is also highly developed in Italy, where the term *dietrologia* ("behindology") describes the rich trove of stories about what lies "behind appearances."

16. Foucauldian-inspired studies of governmentality typically fail to pay serious attention to Foucault's emphasis that governmentality is about "governing the relation between men and things" and to consider precisely how it is that things get made. In so doing, materiality disappears with nature, and bureaucratic logics come to seem all powerful, hermetic, and closed.

Chapter 2

1. For further details, see Mathews (2004) and the massive survey of Mexican environmental history by Antony Challenger (Challenger 1998) (unfortunately this valuable book is not available in English).

2. The dominant account of the Mexican Revolution (1910–1920) is of an agrarian revolt. Although this was an important component, participants' motivations were more complex and varied (see Joseph and Nugent 1994).

3. The Mexican state created two forms of rural community. The first form was the *comunidad*, created by either formally recognizing communities with existing and solid land titles (often indigenous communities with colonial era legal titles) or *restitución* (restitution) to communities that had recently lost their land. Most land awarded by the postrevolutionary agrarian reform process was the second form of community, the *ejido*, which was titled through *dotación* (gift). The glacial pace of the *restitución* process forced many communities to settle for *dotación* as *ejidos* even when they were actually prerevolutionary *comunidades* (Craib 2004:242–252; Joseph 1994a). For the remainder of this book, I will lump *comunidades* and *ejidos* together as "communities," although there are legal differences (Turner and Taylor 2003).

4. By 1940, around 18% of forest land was in community or *ejido* hands (Klooster 2003a).

5. Estimates of forest area in Mexico are confused by the fact that "forest land" is land over which the forest service claims jurisdiction because it is supposed to properly support trees, although much of this "forest" is actually grassland. The figure of 80% of forest area as communally owned is repeated throughout the literature (Snook 1995; Merino 1997), but there seems to be no primary source for this.

6. In this book, I will refer to the "forest service" to describe the federal institution or institutions responsible for forests. In reality, the agency responsible for forests has migrated across different ministries over the last century, partly due to its institutional and political weakness. See table 2.1 for a list of changing federal forestry institutions over the last century.

7. The drainage project was managed by the British firm of Pearson and completed by 1900.

8. Miguel Angel de Quevedo was a quintessential Porfirian *cientifico*: Educated in France, he worked on the drainage of the Valley of Mexico and was associated with the limited forestry regulations of the late Porfiriato period. He was active in the intellectual ferment that surrounded the constitution of 1917 (Quevedo 1916) and tried to ensure that agrarian reform would produce communally owned forests that would be regulated by the state (see also Boyer 2005; Simonian 1995).

9. I have been unable to find staffing levels for 1934–1939, but Manuel Hinojosa Ortíz (1958:31–32) gives the following figures for 1953: 223 foresters, 464 administrators, 1,093 forest guards, and 1,200 laborers in tree nurseries. These numbers agree with informants' estimates of staffing levels in 1938–1940.

10. Most of these concessions were in temperate pine forests in the Sierra Madre Oriental and Occidental in the states of Michoacán, Jalisco, Durango, Chihuahua, and Oaxaca (Mejía Fernández 1988; Huguet 1953). Tropical forests in Chiapas and Yucatan had been largely logged of large trees during the late nineteenth and early twentieth centuries, but in the 1940s, forest concessions were awarded in the Yucatan peninsula in forests previously economically inaccessible due to lack of streams to float logs out (Klepeis and Turner 2001; Haenn 2002).

11. By 1967, 200 out of the 700 forest guards working in the whole country were employed by UIEF (SAG 1967; Anzures Espinosa 1967). The logging bans were enforced by these guards, and between 1946 and 1960, by a militarized federal forest police, the *Policía Federal Forestal*.

12. The UIEF were initially privately owned companies that allowed the participation of foreign investors in association with national political elites. President Miguel Alemán (1946–1952) was involved with the FAPATUX logging concession in Oaxaca through his wife, who was associated with the Canadian investor George Wise (FAPATUX 1956). Alemán was also directly involved with the *Caobas Mexicanas* logging concession in the state of Campeche (Bray and Klepeis 2005).

13. By the end of the 1970s, more than 2–3% of national forest production came from *ejidal* forests (Klooster 2003a).

14. The community forestry sector has been able to press for government subsidies in the last three presidential administrations, including most recently under President Calderon, when an environmental services program was altered to ensure that forest communities were paid for protecting their forests.

15. *Gremio*: "Corporation formed by the teachers, officials and apprentices of a single profession or office, governed by special ordinances and statutes" (Real Academia Española 2003).

16. Note that this map represents the "predisturbance" distribution of the pine oak forest type and may overstate the original forest distribution in order to claim authority for the forest service.

Chapter 3

1. I owe this insight to extended conversations with Anna Tsing, who has her own formulation for this kind of movement.

2. The northern side of the Sierra Juárez receives high rainfall, with annual precipitation ranging from 2,000 mm in the Valle Nacional to 6,000 mm at the top of the range, while the Valley of Oaxaca is semi-arid, receiving only around 600–800 mm per year. The higher the elevation, the longer the rainy season: At the top of the Sierra Juárez, the dry season may only extend from November to February, whereas in the Valley of Oaxaca, the dry season can extend from November to May (INEGI 1980a, 1980b, 1990a, 1990b, 1990c). Average annual temperatures decrease with elevation, falling from 26°C in the Valle Nacional to as little as 11°C at the top of the Sierra Juárez (INEGI 1980b).

3. Between 1,400 and 2,200 meters above sea level.

4. Annual precipitations in this forest type can range up to 4,500 mm, although 2,000–3,000 mm is more usual (INEGI 1990a, 1980a).

5. The dry season lasts five to six months and rainfall ranges from 1,200 to 2,000 mm per year (INEGI 1990a).

6. This outline is drawn from Rzedowski (1978a, 1978b) and Lorence and García Mendoza (1989). For an exhaustive discussion of the natural history of Mexico, see Antony Challenger's magnificent work (Challenger 1998).

7. For a thorough discussion of the impact of the conquest on the Sierra Juárez, see Chance (1989).

8. A *mestizo* is literally someone of combined Spanish and indigenous descent. In practice, being *mestizo* is as much a cultural performance as a biological fact. There is vast literature on *mestizaje*. For some initial points of entry, see Beltrán (1979), Smith (1996), and Wade (1997).

9. A *caudillo* is a military/political leader who rules through personal charisma and patron–client relations. These figures were omnipresent in Latin America throughout the nineteenth and early twentieth centuries.

10. The *Porfiriato* is the period of Porfirio Díaz's rule, 1876–1911.

11. Rosendo Pérez García describes communities as cohesive units, in accordance with a *serrano* tradition of concealing community conflicts from outsiders. Communities were certainly *not* unified in their response to mapping efforts (McNamara 2007; Pérez García 1996a [1956], Pérez García 1996b [1956]).

12. Over the nineteenth century, standards of survey accuracy improved, and rural people came to learn about "scientific" mapping and its associated conventions of graphical representation (Craib 2004).

13. At a time when the population of the district of Ixtlán was around 24,000 (Pérez García 1996a [1956]:435–446), the most important mines at La Natividad, near Ixtlán, employed around 450 workers. Smaller mines at the nearby communities of Yavesia, Lachatao, and Amatlan employed most of the adult men in these towns.

14. Ixtlán, Atepec, Jaltianguis, Guelatao, and Chicomezúchil are all said to have specialized in growing food for sale to the mining communities (Pérez García1996a [1956]:316).

15. Oak species are preferred for charcoal production and for fuelwood where sustained high temperatures are required.

16. For example, Benito Mussolini was named after Benito Juárez.

17. Contests over the memory of Juárez and Díaz were critical to relationships between the federal government and its former military supporters in the Sierra (see McNamara 2007).

18. Environmental historians and political ecologists have increasingly made use of landscape photographs to interrogate landscape histories. For an excellent assessment of these methods, see Kull (2005).

19. This is an often repeated point that nevertheless bears further repeating. For an excellent example, see Paul Robbins' comparison of forester and pastoralist readings of landscape in Rajasthan, India (Robbins 1998, 2003).

20. Although Hérnandez and Meixueiro did try to gain control of some land, their overall holdings were relatively modest (McNamara 2007).

21. This is a vastly complicated period. In the Sierra Juárez, opposing factions allied themselves with national movements but usually for reasons to do with intercommunal conflicts over land. The Maderistas were the supporters of President Francisco Madero, who briefly supplanted Díaz from 1911 to 1913. The constitutionalists were the followers of President Venustiano Carranza, who succeeded in defeating both federalists (followers of the Díaz regime) and the agrarian reformers Emiliano Zapata and Pancho Villa. Carranza was defeated and killed in 1920, to be replaced by Álvaro Obregón, who was assassinated in 1928.

22. It is not clear what degree of damage Ibarra means by "burned"; in the case of Ixtlán, municipal records were burned and many of the cattle and sheep were stolen or killed. In the case of Ixtepeji, the town was razed to the ground and the church bells and altarpieces were taken away. Other communities said to have been "burned" include Xopa, Guiloxi, Yahuio, San Francisco Cajonos, San Pedro Cajonos, and San Miguel Cajonos.

23. A *comunero* is an adult, usually male, community member who has political and economic rights over communally owned lands. If the "community" is legally an *ejido*, he is called an *ejidatario*.

24. There were invasions and counterinvasions by different factions, including an attempt by the serranos to control the entire state in 1915, followed by an incursion into the Sierra by the constitutionalist forces of Carranza in 1916.

25. By 1916, people from Ixtepeji were allied with the constitutionalist faction of president Carranza.

26. Young (1982) gives a figure of one truck replacing thirty-three mules or eighty donkeys and from eight to eighteen men engaged in the transport business.

27. Mobile landscapes and species can confound state or scientific efforts to fix plants in place (Raffles 2002; Tsing 2005). Fire, as a powerful agent of landscape transformation, opens up the possibility of these kinds of mobilities.

28. I have used the ecological-type system of Rzedowski (Rzedowski 1978a, 1978b).

29. See also Fulé and Covington (1999); Fulé, Villanueva-Díaz, and Ramos-Gómez (2005); Heyerdahl and Alvarado (2003); and Segura and Snook (1992).

30. The limited ecological research on the pine and pine-oak forests of Mexico confirms that Mexican pines are drought-tolerant, shade-intolerant pioneer species that can tolerate poor sites and dominate stands in earlier successional stages (Asbjornsen 1999), whereas oaks can also tolerate poor sites but tend to be mid-successional and are able to tolerate partial shade (González-Espinosa et al. 1991; Quintana-Ascencio et al. 1992).

31. In parts of the world where disturbances are sufficiently rare, oaks are in turn suppressed by still more shade-tolerant species, such as sugar maples and beeches in New England.

32. Total district population in 1946 was 33,000, when Ixtlán had 1,165 inhabitants (Pérez García1996a [1956]:435–446). This would yield an agricultural area of around 840 hectares in 1946, the year nearest to the first agricultural census in 1950 (Abardia Morelos and Solano Solano 1995).

Chapter 4

This chapter draws on the *Archivo General del Estado de Oaxaca* (AGEO), which contains forestry disputes concerning the governor between 1922 and 1956. Contemporary newspaper accounts are from the state newspaper archive, the *Hemeroteca Estatal de Oaxaca* (HEO). Legal disputes about forests take place in the federal court system and are not open for research.

1. For more detail on Cárdenas's visit to Oaxaca and his negotiation with *serranos*, see Smith (2002).

2. The PNR (*Partido Nacional Revolucionario*) became the PRM (*Partido Revolucionario Mexicano*) in 1938 and then the PRI (*Partido Revolucionario Institucional*) in 1946.

3. For a full description of the *soberanista* movement, see Garner (1988) and Smith (2008).

4. An *ixtleco* is someone from the community of Ixtlán de Juárez.

5. The *cargo* system is the traditional system of office holding within indigenous communities in Mexico and elsewhere in Latin America. Male community members hold a series of public offices, of increasing responsibility and authority, ideally progressing through all or most of the levels of the hierarchy, and ending up as *ancianos* (elders), who have great power over decision making. The *cargo* system draws many of its job titles from colonial Spanish legal tradition and has been the subject of an enormous literature, which variously portray it as a mechanism for equalizing community wealth (Wolf 1957, 1986) or of intracommunity domination and abuse (Rus 1994). The cargo system has taken different forms at different places and times and has been greatly impacted by interactions with different national and regional legal regimes.

6. National laws could take a long time to penetrate to the Sierra: Although tithes were formally abolished in 1833, they were still being paid in a community near Ixtlán in 1933 (Young 1975:251).

7. For example, with remnants of pre-Colombian settlements, which are scattered throughout the Sierra.

8. De la Fuente says "*en los cerros*," which may have meant specific places in the mountains, as opposed to mountain tops proper.

9. This association between water and the spirits of mountains, springs and caves continues to be widespread across central Mexico (Glockner 1999; Goloubinoff, Katz, and Lammel 1997; Hémond and Goloubinoff 1997; Villela and Samuel 1997), while weather prediction practices are apparently examples of syncretism of Spanish and pre-Hispanic beliefs about climate.

10. This statement resonates with a study by Lazos and Paré (2000), who found that older *Nahua* indigenous people in a community in Veracruz were more likely to believe that degradation was the result of improper relations with nature spirits, whereas younger people were more likely to attribute degradation to deforestation and environmental degradation.

11. Manuel Hinojosa Ortíz, a former head of the forest service, stated that in 1953 the national forest service employed 223 technicians, 464 administrators, and 1,093 police, of whom half were based in Mexico City (Hinojosa Ortiz 1958:31–32). This would suggest that in Oaxaca in 1953, there would have been roughly four forestry technicians, eight administrators, and eighteen forest police. Oaxaca, with a relatively large area of forest, may have had more staff than this calculation suggests, but even so this was not many people to police 1 million hectares of pine forest.

12. This is dramatically different from the situation of the numerous forest guards employed at the lowest levels of the Indian forest service, and their much greater degree of involvement in community life (Robbins 2000).

13. Communities in Oaxaca often have multigenerational and even century-long land disputes with their neighbors (Dennis 1987).

14. There is a rich tradition of studying state rituals in Mexico. For a recent summary, see Lomnitz (1995).

15. The railway line was completed in 1897 and used firewood to fuel the trains until well into the twentieth century. The impact of the railway on forests is mentioned by Conzatti in 1913 (Bradomín 1953d).

16. Using modern-day estimates of firewood use of 0.81 cubic meters per person per year (Masera et al. 1997) and the population figures for 1950 of 1.4 million people in the state of Oaxaca (García 1996) gives a total fuelwood consumption (for firewood, charcoal, brick-making, etc.) of approximately 1.1 million cubic meters per year.

Chapter 5

1. According to admittedly self-serving documents, at the height of FAPATUX operations in the mid-1970s, the company directly employed more than 150 full-time and 3,500 part-time workers in a relatively small area of forest (Tamayo, 1976).

2. For an example of a similar negotiation of new forms of knowledge and political power, see the shifting epistemic, discursive, and political negotiations surrounding the listing and delisting of African elephants from CITES (Thompson 2004).

3. The percentage of Zapotec speakers has declined steadily over the last fifty years (Varese 1983). In Ixtlán there are only a few older Zapotec speakers. In coffee-growing communities only a few hours' drive away, most people still spoke Zapotec by preference when I visited in 2001.

4. The Papaloapan river watershed encompasses the entire Sierra Juárez and parts of the adjacent Mixteca.

5. An analogous example is the case of the floods that took place in Thailand in late 1988, attracting international media attention and triggering a reevaluation of national forest policies. For more on the history of forests and watersheds in Thailand, see Forsyth and Walker (2008).

6. In a typical example of such a contract from 1962, FAPATUX offered a direct payment to the community of Nuevo Zoquiapam of $5.00 per cubic meter, an additional payment of $15.00 per cubic meter to the *Banco de Credito Ejidal* (Ejidal Credit Bank), and established the price it would pay community loggers (AGEO Various 1963).

7. In fact, as we have seen in our discussion of pine ecology in chapter 4, these apparently "young" trees would mostly have been suppressed and would not have grown up to become the dominant trees of the future. On the contrary, although the scarification of mineral soil by clear-cut logging appeared devastating, a properly applied clear cut would have produced the necessary seedbed to ensure prolific pine regeneration (Snook 1986). Foresters use term "clear cut" in a specific technical sense to describe the total removal of the standing crop of

trees, allowing the regeneration of disturbance-adapted species such as pines. FAPATUX logging would almost certainly have been what is called a "commercial clear cut" that removed commercially attractive trees and left behind damaged or defective trees, gravely damaging the future value of the forest. For further explanation of these terms, see Smith (1997).

8. This change in logging methods and the associated reduction in profits may have been one of the causes of the withdrawal of foreign investors in 1965, when FAPATUX was nationalized.

9. Forestry technicians were FAPATUX employees, as were the *documentadores,* (documenters), the *recibidores* (receivers), and the *jefe de monte* (forestry crew chief). The *documentador* was responsible for producing the detailed manifests required to transport timber, and the *recibidor* was responsible for periodically measuring the timber stacked at the roadside. Another important group of workers were the bulldozer drivers and technicians who laid out new logging roads. Finally, foresters and forestry technicians were responsible for locating timber and for marking trees to be cut, and the director of forestry was responsible for negotiating logging agreements.

10. Each two-man logging crew cut perhaps 10 cubic meters per week.

11. These estimates agree with Jorge L. Tamayo, the director of FAPATUX in the late 1970s, although Tamayo did not give precise figures for the size of the "very large" trees cut (Tamayo 1976).

12. FAPATUX gradually expanded saw milling operations during the course of the concession; by 1978, around 45% of timber was processed at saw mills in Etla, Ixtlán, and Concepción Papalos, and the rest was turned into newsprint in Tuxtepec (FAPATUX 1977).

13. Modern estimates of firewood use of 0.81 cubic meters per person per year and a 1980 population of 60,000 people for the districts of Ixtlán and Villa Alta (García 1996) give a fuelwood consumption of approximately 48,000 cubic meters. For the state of Oaxaca, consumption of around 2.4 million cubic meters in 1980 would have been several times greater than industrial timber production of 500,000 cubic meters. Estimates of fuelwood use are notoriously imprecise as users vary their consumption depending upon available technology, scarcity, cost, and cultural considerations (Donovan 1981).

14. Similarly, paternalistic *ejido* and community forestry businesses were being tried in other parts of Mexico at this time (Enriquez Quintana 1976).

15. *Organización para Defensa de los Recursos Naturales y Desarrollo Social de la Sierra de Juárez* (Organization for Defense of the Natural Resources and for the Social Development of the Sierra Juárez) (Mecinas and Martinez Luna 1981; Martinez Luna 1977).

16. One such community is Guelatao, about a mile from Ixtlán. As the birthplace of Benito Juarez, Guelatao has had a long history of contact with state indigenous institutions, including the presence of a state boarding school for indigenous people and an INI office.

17. In some communities logging officially ceased for several years, but in at least some of these cases logging continued illegally.

18. Catherine Young (Young 1982) gives migration rates for the district of Ixtlán of 246 per 1,000 between 1950 and 1960, and a decline of 2.3%, in the population. Rural population stability was only maintained by exporting young people of working age.

19. I am indebted for this observation to Michael R. Dove.

20. In much of Mexico, there is a strong tradition that converting forest to a swidden field makes it the *de facto* private property of the farmer.

21. Zenaido's use of the term "irrational exploitation" appropriates official language of forest management, found in the original FAPATUX concession (Presidencia de la República 1958a), as in present-day forest service publications (Fonseca et al. 2000).

22. See, for example, the way that swidden agriculturalists in South East Asia adopted the cultivation of Brazilian rubber (*Hevea brazilensis)* during the nineteenth century (Dove 1994).

Chapter 6

1. Portions of this chapter have appeared in Mathews (2005, 2009). My discussion owes much to years of conversations with Michael Dove and David Graeber, although my formulation of power/ignorance (Mathews 2005) is rather different from that of Graeber (2006).

2. The power to bore is perhaps one of the most characteristic forms of bureaucratic power, the ability to immobilize the imagination. Bureaucrats can make you fill in a form that tells nothing of who you are, and they impose boredom on you as you wait in line for important stamps and signatures.

3. See, for example, Alain Desrosières' discussion of the ways in which forms of statistics are linked to scales, institutions, and interventions on society (Desrosières 1991).

4. Governor Jose Murat Casab had his own projects for conservation/development, focused on the construction of a transisthmus highway, which was also supposed to protect tropical nature.

5. This section is drawn from seventeen extended interviews with present or former SEMARNAP officials in Mexico City in 1998, 2000–2001, and an additional seventeen interviews in 2008 and 2009 on a related research project on climate change policy. See the appendix for a list of interviews. Navigating the bureaucracy as I did, I was never firmly ensconced in one place. I never got to sit in an office and watch the flow of documents and the daily give and take of an office, as in classic sociological accounts of bureaucracy and science. Certainly, I often felt nostalgic for a fixed field site where I could have gotten to know the jokes, the people, and the life histories, and I supplemented these interviews with analyses of official reports and documents.

6. The *pirul* (*Schinus molle*), also known as the Brazilian pepper tree, is a common urban street tree in Mexico.

7. Antonio is a well-known legal sociologist who has written books on such subjects as the relationship between formal legal systems and the practices of urban squatters (Azuela et al. 1998; Azuela and Tomas 1997; Azuela 1995; Chirinos and Azuela 1993).

8. Formerly, many senior positions were filled by politicians trained in law or economics at the UNAM. With the declining prestige of the Mexican state, politicians have tried to increase the legitimacy of their administrative appointments by filling senior positions with technocrats or scientists (Lomnitz, Mayer, and Rees 1983).

9. The PRI is the Institutional Revolutionary Party, the PAN is the National Action Party, and the PRD is the Party of Democratic Revolution. The transfer of power to the PAN in 2000 marked a seismic shift in Mexican politics, as the PRI had maintained control of the presidency since 1929.

10. Many people had hoped that the new Fox administration would bring corruption charges against former officials. Perhaps because of his party's lack of a parliamentary majority and his need for support from the PRI, Fox had little interest in bringing former officials to justice.

11. The original reads: "Yo no sé que es 'sustentable'/ pero hay que ponerle acento/ por lo pronto es muy amable/ porque a mí me da el sustento." Although open criticism was traditionally avoided in Mexican politics, there is a tradition of expressing dissent through barbed poems and epigrams that appear in newspapers or in extemporaneous performances in public events. This tradition was at its height in the 1950s and 1960s but persists to this day.

12. The *tlacuache* is an indigenous loan word for the common opossum. This usage suggests that career officials, like nocturnal opossums, make a living by scrambling around in the shadows of power and perhaps that a good bureaucrat knows when to defer to power by playing dead. Mexican Spanish contains numerous indigenous loan words that are commonly used in discussing politics, indicating perhaps the perception that these things are particular to Mexico or to its indigenous past and cannot be expressed in terms known to outsiders. Political figures may be called *caciques* (indigenous chiefs), the president may be ironically termed the *huey tlatoani* (a term for the Aztec emperor), and so on.

13. Reforestation budgets are often dedicated to other ends by local-level officials (personal communication, Nora Haenn, December 1, 2005). See also Haenn (2005:156).

14. This practice is not confined to Mexico: Monique Nuijten tells me that anonymous memos also circulate in the Danish Foreign Ministry (personal communication, 2006).

15. The forest service has been reorganized seventeen times over the last century (see table 2.1 in this book).

16. For reasons of space, I can no more than suggest the bodily nature of this performance. To describe such performances as merely rhetorical is to lose track

of what Michael Herzfeld calls "social poetics" (Herzfeld 1997:139–155) and to overemphasize the linguistic and the structural at the expense of materiality, causation, and performance.

Chapter 7

1. SEMARNAP lost responsibility for fisheries in 2001, becoming the ministry of Environment and Natural Resources (SEMARNAT). Responsibility for forest subsidies, reforestation, and extension were transferred to the semi-autonomous National Forestry Commission (CONAFOR) under the prominent PAN politician Alberto Cárdenas. In 2009, CONAFOR and SEMARNAT offices were a few blocks from each other in the Reforma neighborhood, and many of the people who had worked in SEMARNAP in 2000–2001 were distributed across the two institutions.

2. Municipalities are in turn divided into *comunidades agrarias* (agrarian communities); for example, the municipality of Ixtlán is divided into the communities of Ixtlán, Yagila, Tepanzacoalco, and so on, each with its own *comisariado de bienes comunales*.

3. Community forest holdings vary greatly in size: an arithmetical average of around 2,800 hectares per community and 23,000 hectares per consulting forester gives a very rough sense of the areas and people concerned.

4. There were regional SEMARNAP offices in Oaxaca, in Ixtlán, at Pinotepa in the Sierra Sur, in Puerto Escondido on the south coast, at Tlaxiaco in the Mixteca, at Ixtepec in the Isthmus of Tehuantepec, and at Tuxtepec in the Valle Nacionál.

5. In 2001, *servicios técnicos*' charges for marking timber varied from M$6.00 per cubic meter to M$20.00 per cubic meter out of a market value of approximately M$850.00 per cubic meter for pine timber delivered to the saw mill (Raul Alvarez Castillo, interview notes, Oaxaca, May 17, 2000). In 2000, total PROCYMAF payments to *servicios técnicos* were M$4 million (DeWalt and Guadarrama Olivera 2000) out of a total of M$10.5 million earned by foresters, assuming that they had earned M$10.00 per cubic meter (Plancarte Barrera 2000).

6. This is what happened in 2002 when the Oaxaca state director of SEMARNAT was accused of issuing a logging permit on contested land in the Sierra Sur of Oaxaca, ultimately causing a massacre (Weiner 2002).

7. A logging operation is considered a "treatment" in the medicalized language of silviculture, which seeks to tend a forest stand from initiation to harvesting.

8. Numerous ethnographers of bureaucracy have pointed to this kind of detachment of bureaucratic standards from substantive practices (e.g., Power 1997; Riles 1998).

9. Total officially declared forest production in Oaxaca in 1999 was around 670,000 cubic meters (Plancarte Barrera 2000). More than 70% of timber production was pine saw timber, with the remaining 20% processed as pulpwood

by the FAPATUX paper factory in Tuxtepec (SEMARNAP 1997a; Gobierno de Oaxaca 2003).

10. Gustavo said, "*hacen un gran discurso*." The closest translation might be, "They talk a good line," but this suggests a vernacular and almost folksy genre, whereas Gustavo was suggesting intimate conversations that also had official, bureaucratic, and political connotations.

11. Political corruption has been relatively understudied by anthropologists, who traditionally left the field to political scientists (e.g., Nye 2002[1967]; Huntingdon 2002[1968]; Scott 1969; Ascher 1993). More recently, there has been a proliferation of ethnographic analyses of corruption (Gupta 1995; Coronil 1997; Lomnitz 1995; Mamdani 1996). Following Gupta and Lomnitz, I see corruption as a cultural category that links states, citizens, and other organizations. I argue that under some circumstances skilled corruption talk can do political work and can make officials accountable to public pressure (see also Olivier de Sardan 1999; Nuijten 2007).

12. This awkward translation conveys a passive and reflexive verb usage that concealed his active decision to ignore the regulation.

13. I owe this usage of "transposition" to Julie Chu, who describes how people work with immigration forms in China and how the act of filling in a form forecloses the details of people's personal lives as they make themselves legible by providing only the kinds of administrative information that the form seeks (Chu 2009).

14. *Usos y costumbres*, literally usages and customs, refers to the state of Oaxaca's official acknowledgment of a separate political and legal system within indigenous communities. See Nader (1990).

15. The FAO is the Food and Agriculture Organization of the United Nations.

Chapter 8

1. The municipality of Ixtlán contains the communities of La Josefina, La Luz, La Palma, San Juan Yagila, San Miguel Tiltepec, Santa Cruz Yagavila, Santa Maria Josaa, Santa Maria Yahuiche, Santa Maria Zoogochi, Santiago Teotlaxco, Santo Domingo Cacalotepec, and San Gaspar Yagalaxi, with a total population of approximately 4,000 people (INEGI 2003).

2. The UCFAS was the *Unidad Comunitaria Forestal Agropecuaria y de Servicios* (the Community Forestry, Agriculture and Services Unit).

3. Someone from Ixtlán.

4. It is tempting to guess that the source for Pozo's knowledge of Nazi era forestry came from the 1940–1942 era publications of a group of conservative anti-American intellectuals led by the immensely influential José Vasconcelos. This group was affiliated with the Catholic conservative *Sinarquista* movement and more loosely with Spanish fascism. Perhaps the pro-Nazi magazine *Timon*, which ran from 1940 to 1942 and was edited by Vasconcelos, contained some description of German forestry practices, and perhaps this journal reached

Pozo. For discussions of Nazi era forestry, see Nelson (2005) and Schama (1995).

5. The *municipio* of Ixtlán contained several other autonomous "communities": In descending order of administrative importance, these are the *comunidad*, *agencia*, and *ranchería*, collectively known as *sujetos* (subjects), of which Ixtlán is the *cabecera* (head).

6. *Cargo* systems of community government have been the subject of intense debate over whether they are redistributive and defensive mechanisms (Wolf 1955, 1957), a means of state control within communities (Rus 1994; Friedlander 1981), or whether community elites manipulate the system to reinforce existing economic inequalities. Recent scholarship has concluded that *cargos* have little leveling effect and can weaken a community by pumping out the money used in fiestas (Chance and Taylor 1985), but there is no consensus about whether it helps or hinders state penetration. This suggests a conceptual confusion: Just as the state is an idea that is used by specific people in order to build institutions or inaugurate development projects, so too the *cargo* system is better seen as a set of institutions that are continually remade using ideas about community, public responsibility, and the state. Like fragmented and unstable state institutions, communities can be well organized or drastically disorganized, lacking regular meetings, members, budgets, archives, and officeholders (Nuijten 2003).

7. For a classic discussion of the relationship between community and federal legal systems, see Nader (1990).

8. In 2000, the distribution was more than three months of the minimum wage for UCFAS employees.

9. In 2000, the *sindico* was paid M$800 per month, and the minimum wage for UCFAS employees was about M$1,250 per month.

10. Laura Nader has written about the ideology of harmony in dispute resolution in the nearby community of Talea de Castro (Nader 1990).

11. The cost of the plan was M$200,000, around US$20,000.

12. Anna Tsing has pointed out to me how the technical language of forestry is highly medicalized. Logging is a "treatment." Foresters speak of logging "operations," of "tending" stands, and so on. This suggests a past interdisciplinary borrowing of high-status medical terms by a forestry profession that wanted to claim professional authority and care.

13. The philosophical basis of forestry has changed greatly over the last two 200 years, so this is necessarily the briefest of summaries. For an introduction to silviculture, see Smith (1997).

14. This is similar to the concept of "torque" described by Geoffrey Bowker and Susan Leigh Star, as when theories about tuberculosis and the bodies of diseased people change through time (Bowker and Star 1999).

15. See Scott (1998). Critically, and rather contrary to the ways that Scott is usually interpreted, he emphasizes that projects of legibility seek to *remake* the landscape, rather than being an apparently simple act of reading though inspection.

16. For a full discussion of the politics of forests and waters in Ixtlán and in Mexico, see Mathews (2009).

17. Actor-network theory (ANT) has been criticized by a number of theorists, including Latour himself (Fischer 2007), but the principal insight of this approach, that knowledge is made by "the careful plaiting of weak ties," remains valid. In addition to speaking of networks, I have therefore used terms such as "weaving," "entanglement," and "braiding," all of which refer to performance, to possible remakings, and to nodes rather than solely to networks.

References

Abardía Morelos, Francisco, and Carlos Solano Solano. 1995. *Empresas forestales comunitarias en las Américas: Estudios de caso*. Paper presented at *Aprovechamientos Forestales en las Américas: Manejo Comunitario y Sostenibilidad*. 3–4 February, University of Wisconsin–Madison.

Agee, James K. 1998. Fire and pine ecosystems. In *Ecology and biogeography of Pinus*, ed. D. M. Richardson. Cambridge: Cambridge University Press.

AGEO Various. 1963. Letter from FAPATUX director Andrés Benton Cuellar to Presidente Municipal of Zoquiapan. AGEO Permisos de aprovechamiento forestal: Asuntos Agrarios.

Agrawal, Arun. 2001. Common property institutions and sustainable governance of resources. *World Development* 29 (10):1649–1672.

Agrawal, Arun. 2005. *Environmentality: Technologies of government and the making of subjects*. Durham: Duke University Press.

Aguirre Beltrán, Gonzalo. 1979. *Regions of refuge*. Washington, DC: Society for Applied Anthropology.

Alcorn, Janis B. 1984. *Huastec Maya ethnobotany*. Austin: Texas University Press.

Altamirano Piolle, Maria Elena. 1993. Entre las montañas de Oaxaca. In *José Maria Velasco: Paisajes de luz, horizontes de modernidad*. Mexico City: Museo Nacional de Arte.

Anonymous. 1922a. *El Jefe del sector forestal en el estado dice que queda definitivamente instalada oficina forestal*. AGEO: Asuntos Agrarios, Serie V, Problemas por Bosques, Legajo 892, Expediente 24.

Anonymous. 1922b. *Relativo a la queja en contra del citado lugar por ir al bosque a cortar leña y hacer carbon*. AGEO: Asuntos Agrarios, Serie V, Problemas por Bosques. Legajo 892, Expediente 7.

Anonymous. 1929. *Se comunica al delegado forestal que los vecinos de San Pablo Etla y San Felipe del Agua están talando los montes indebidamente*. AGEO: Asuntos Agrarios, Serie V, Problemas Por Bosques, Legajo 892, Expediente 14.

Anonymous. 1930. *Suplica se sirva dictar las ordenes conducentes a evitar la introduccion de productos forestales fraudulentos.* AGEO: Asuntos Agrarios. Problemas Por Bosques, Legajo 893, Expediente 2.

Anonymous. 1931a. *Acta constitutiva y estatutos de la sociedad cooperativa forestal de.* AGEO: Asuntos Agrarios, Problemas Por Bosques, Legajo 892, Expediente 17.

Anonymous. 1931b. *El Jefe de la Sección Forestal y de Caza y Pesca denuncia que varias personas detienen madera, haciendose pagar como empleados.* AGEO: Asuntos Agrarios, Problemas Por Bosques, Legajo 893, Expediente 13.

Anonymous. 1931c. *El Presidente Municipal de San Felipe del Agua se queja de que el C. Inspector Forestal en esta Ciudad no deja vender su leña a los vecinos de ese municipio.* AGEO: Asuntos Agrarios, Problemas Por Bosques, Legajo 892, Expediente 17.

Anonymous. 1932. *El presidente municipal de San Felipe del Agua se queja de que el C. Inspector Forestal en esta ciudad no deja vender su leña a los vecinos de ese mun*icipio. AGEO: Asuntos Agrarios, Problemas Por Bosques, Legajo 893, Expediente 18.

Anonymous. 1942. *Los vecinos de Ixtlan de Juarez denuncian que los rancheros de Rancho Chivo han provocado un incendio.* AGEO: Asuntos Agrarios, Problemas Por Bosques, Legajo 900, Expediente 18.

Anonymous. 1947. Editorial. La Voz de Oaxaca 5 (27):3.

Anonymous. 1949a. Oaxaca y su reforestación. La Voz de Oaxaca 11 (17):2.

Anonymous. 1949b. Testimonio de amor al Arbol, bajo las frondas del sabino del Tule. *La Provincia* 29 (3):2.

Anonymous. 1950. Marzo y Abril: Improprios para hacer plantaciones. *La Provincia* 27 (3).

Anonymous. 1950–1980. Documents about land disputes between Yavesia, Lachatao, and Amatlan in Pueblos Mancomunados, Oaxaca, Mexico. Archivo del Ex Cuerpo Consultivo Agrario.

Anonymous. 1951a. Absoluto respalde presidencial para contener la deforestacion: asi fue garantizado al Gobernador Mayoral Heredía, durante reciente entrevista que tuvo con el Licenciado Miguel Alemán. *La Provincia* 12 (1):1–2.

Anonymous. 1951b. El Juez de Miahuatlan Lic. Miguel Figueroa fue destituido y consignado a la Procuraduría: Fue acusado por el tribunal de abuso de autoridad pues resulta responsable de que los madereros Ranz hayan obtenido su libertad. *Nuevo Diario* 13 (7):1.

Anonymous. 1951c. Se toman medidas para investigar si cumplen los madereros con el convenio que firmaron. *Nuevo Diario* 10 (30):1, 4.

Anonymous. 1951d. Ya se encuentran presos los talamontes Ranz Iglesias. *La Provincia* 21 (1):1–2.

Anonymous. 1952. Oaxaca viste de luto y dolor!! Anoche corrio sangre de la masa popular que protestaba en contra de la implantacion del codigo fiscal. El Imparcial 3 (21):1.

Anonymous. 1953a. Se cortara de raíz la inicua explotación de los rapamontes en nuestro estado: habran de poner coto a su insaciable ambición. El Imparcial 8 (12):1.

Anonymous. 1953b. La carretera a Guelatao y la estatua a Juárez. El Imparcial 5 (14):2.

Anonymous. 1956. En la cuenca del Papaloapan se desarrollan obras de gran importancia. *El Imparcial* 10 (21):2–3.

Anonymous. 1954. Gran entusiasmo reina en los pueblos de la Sierra Juárez: Ya va a iniciarse la construcción de la carretera Oaxaca-Guelatao. El Imparcial 5 (20):2.

Anonymous. 1958a. Nuevo peligro de un monopolio forestal: los campesinos oaxaqueños tratan de evitar esta inicua explotación de riqueza forestal de Oaxaca. *El Imparcial* 9 (11):1, 6.

Anonymous. 1958b. Criminal proceder de talamontes para con campesinos de Cajonos: Despues de hacerlos ingerir bebidas embriagantes los hacen firmar documentos perjudiciales. *El Imparcial* 9 (1):1, 4.

Anonymous. 1959. Continua la inmoderada tala de arboles en pueblos de Cajonos: Una empresa maderera se ha puesto de acuerdo con las autoridades municipales del lugar. *El Imparcial* 4 (1):1.

Anonymous. 1983a. Carta abierta al C. Lic. Miguel de la Madrid Hurtado, presidente constitucional de la republica. Al C. Lic. Pedro Vazquez Colmenares gobernador constitucional del estado de Oaxaca. *Noticias*, February 18, 1983.

Anonymous. 1983b. *La Lucha Comunal en Defensa del Monte: Y nuestro bosque*? Oaxaca.

Anonymous. 1983c. Piden a MMH cancele concesiones a Pandal y FAPATUX. La misma petición para el gobernador del estado. *Carteles del Sur*, February 24, 1983.

Anonymous. 1983d. Respuesta coordinada en Oaxaca: Contra la tala irracional de los bosques. Mas de 20 comunidades forestales de Oaxaca se han amparado colectivamente contra el regalo Lopez Portillista a Pandal Graff Y FAPATUX. *Hora Cero*, March 9, 1983.

Anta Fonseca, Salvador, and Juan Manuel Barrera. 2000. Alternativas en el desarrollo forestal comunitario. *Noticias: Suplemento Ecologia* 10 (21):1–2.

Anta Fonseca, Salvador, Antonio Plancarte Barrera, and Juan Manuel Barrera Terán. 2000. *Conservación y manejo comunitario de los recursos forestales en Oaxaca*. Oaxaca: SEMARNAP.

Anzures Espinosa, Ruben. 1967. El Fondo Nacional de Fomento Ejidal. *Revista del México Agrario* 1 (1):59–69.

Appuhn, Karl. 2000. Inventing nature: Forests, forestry, and state power in Renaissance Venice. *Journal of Modern History* 72:861–889.

Arce, Alberto, and Norman Long. 1993. Bridging two worlds: An ethnography of bureaucrat-peasant relations in western Mexico. In *An anthropological critique of development: the growth of ignorance*, ed. M. Hobart. London: Routledge.

Arno, Stephen M., and Kathy M. Sneck. 1977. A method for determining fire history in coniferous forests of the mountain west. In *USDA Forest Service General Technical Report INT-42*. Ogden, Utah: Intermountain Forest and Range Experiment Station, Forest Service, USDA.

Arteaga, Baldemar. 2000. Enseñanza forestal en México. *Forestal* XXI:21–23.

Arteaga, Baldemar. 2001a. Graph of graduates in forestry from Chapingo 1930–1998.

Arteaga, Baldemar. 2001b. *Situación de la ensenanza e investigación forestales en Mexico*. Texcoco: Universidad Autonoma de Chapingo.

Asbjornsen, Heidi. 1999. Ecological consequences of habitat fragmentation in a human-dominated high-elevation oak-pine forest in Oaxaca, Mexico. Ph.D. Thesis, Yale School of Forestry and Environmental Studies, Yale University, New Haven, CT.

Ascher, William. 1993. *Political economy and problematic forestry policies in Indonesia: Obstacles to incorporating sound economics and science*. Durham, NC: Center for International Development Research, Duke University.

ASETECO and COCOEFO. 2001. Open Letter to Mexican Congress from the Coordinadora de Comunidades y Ejidos Forestales del Estado de Oaxaca. Oaxaca, Mexico. April 2, 2001.

Azuela, Antonio. 1995. *PEMEX: Ambiente y energía, los retos del futuro*. Mexico: Universidad Nacional Autónoma de México, Petróleos Mexicanos.

Azuela, Antonio, Emilio Duhau, and Enrique Ortíz. 1998. *Evictions and the right to housing: Experience from Canada, Chile, the Dominican Republic, South Africa, and South Korea*. Ottawa: International Development Research Centre.

Azuela, Antonio, and François Tomas. 1997. *El acceso de los pobres al suelo urbano*. Mexico: Centro de Estudios Mexicanos y Centroamericanos, Universidad Nacional Autónoma de México-Instituto de Investigaciones, Sociales, Programa Universitario de Estudios sobre la Ciudad.

Barad, Karen. 2007. *Meeting the universe halfway: Quantum physics and the entanglement of matter and meaning*. Durham, NC: Duke University Press.

Barriguete, Manuel V. 1941. *El problema del carbon en México*. Tesis de Ingeniero Agronomo, Bosques, Escuela Nacional de Agricultura, Chapingo.

Bartolome, Miguel Alberto, and Alicia Mabel Barabas. 1990. *La presa Cerro de Oro y el Ingeniero el Gran Dios: relocalización y etnocidio chinanteco en México*. 2 vols. Vol. 1: Dirección general de publicaciones del CONACULTA, Instituto Nacional Indigenista.

Bartra, Armando. 1996. *El México bárbaro: Plantaciones y monterías del sureste durante el porfiriato*. 1st ed. Mexico: El Atajo.

Bebbington, A. 2004. NGOs and uneven development: geographies of development intervention. *Progress in Human Geography* 28 (6):725–745.

Blauert, Jutta, and Kristina Dietz. 2004. *Of dreams and shadows: Seeking change for the institutionalisation of participation for natural resource management. The case of the Mexican regional sustainable development programme (PRODERS)*. London: IIED.

Borgo, Gumersindo. 1998. *México forestal: visto por trece profesionales del ramo*. Mexico: Morelia.

Bowker, Geoffrey C., and Susan Leigh Star. 1999. *Sorting things out: Classification and its consequences*. Cambridge, MA: MIT Press.

Boyer, Christopher R. 2005. Contested terrain: Forestry regulations and community responses in northeastern Michoacán, 1940–2000. In *The community forests of Mexico*, ed. D. B. Bray, L. Merino-Pérez, and D. Barry. Austin: University of Texas Press.

Boyer, Christopher. 2007. Revolución y Paternalismo Ecologico: Miguel Angel de Quevedo y la Politica Forestal en Mexico 1926–1940. *Historia Mexicana* 57 (1):91–138.

Bradomín, José Maria. 1953a. Contra el crimen: IX Urgencia de impedir la deforestación. El Imparcial 4 (8):2.

Bradomín, José Maria. 1953b. Contra el crimen: X Panorama forestal de Oaxaca. El Imparcial 4 (10):2.

Bradomín, José Maria. 1953c. Contra el crimen: XI El mal de México. El Imparcial 4 (13):3.

Bradomín, José Maria. 1953d. Contra el crimen: IX Exposicion cientifica de un sabio. *El Imparcial* 4 (9):1, 4.

Bray, David Barton. 1991. The struggle for the forest: Conservation and development in the Sierra Juárez. *Grassroots Development* 15 (3):13–25.

Bray, David Barton. 1996. Of land tenure, forests, and water: The impact of the reforms to Article 27 on the Mexican environment. In *Reforming Mexico's agrarian reform*, ed. L. Randall. New York: M.E. Sharpe.

Bray, David Barton, and Peter Klepeis. 2005. Deforestation, forest transitions, and institutions for sustainability in south-eastern Mexico, 1900–2000. *Environmental History* 11:195–203.

Bray, David Barton, Leticia Merino-Pérez, Patricia Negreros-Castillo, Gerardo Segura-Warnholtz, Juan Manuel Torres-Rojo, and Henricus F.M. Vester. 2003. Mexico's community managed forests as a global model for sustainable landscapes. *Conservation Biology* 17 (3):672–677.

Bray, David Barton, Leticia Merino-Pérez, and Deborah Barry, eds. 2005. *The community forests of Mexico*. Austin: University of Texas Press.

Brenneis, Donald. 2006. Reforming promise. In *Documents: Artifacts of modern knowledge*, ed. A. Riles. Ann Arbor: University of Michigan Press.

Brosius, J. Peter. 2006. What counts as local knowledge in global environmental assessments and conventions? In *Bridging scales and knowledge systems: Concepts and applications in ecosystem assessment*, ed. Walter Reid, Fikret Berkes, D. Capistrano, and T. Wilbanks. Washington, DC: Island Press.

Brosius, J. P. 1997. Endangered forest, endangered people: Environmentalist representations of indigenous knowledge. *Human Ecology* 25:47–69.

Bruijnzeel, L. A. 2004. Hydrological functions of tropical forests: Not seeing the soil for the trees? *Agriculture Ecosystems & Environment* 104 (1):185–228.

Bryant, Raymond L. 1996. Romancing colonial forestry: The discourse of "forestry as progress" in British Burma. *Geographical Journal* 162 (2):169–178.

Bumpus, A. G., and D. M. Liverman. 2008. Accumulation by decarbonization and the governance of carbon offsets. *Economic Geography* 84 (2):127–155.

Buschbacher, R. J. 1990. Natural forest management in the humid tropics: Ecological, social, and economic considerations. *Ambio* 19 (5):253–258.

Calder, Ian, and Bruce Aylward. 2006. Forest and floods: Moving to an evidence-based approach to watershed and integrated flood management. *Water International* 31 (1):87–99.

Callon, Michel. 1986. Some elements of a sociology of translation: Domestication of scallops and the fishermen of St. Brieuc Bay. In *Power, action, and belief: A new sociology of knowledge?*, ed. J. Law. London: Routledge and Kegan Paul.

Calva Téllez, José Luis, Fernando Paz Gonzalez, Omar Wicab Gutierrez, and Javier Camas Reyes. 1989. *Economia politica de la explotacion forestal en México: Bibliografía comentada. 1930–1984*. Mexico City: Universidad Autonoma de Chapingo, Universidad Nacional Autonoma de México.

Calva Téllez, José Luis. 1997. *El campo mexicano: Ajuste neoliberal y alternativas*. Mexico City: Juan Pablos.

Cárdenas, Lázaro. 1935. Mensaje del C. Presidente de la República, General Lázaro Cárdenas, Radiado al Pueblo Méxicano el 1o de enero de 1935, en lo Concerniente a la Creación del Departamento Autónomo Forestal y de Caza y Pesca. *México Forestal* 13 (January–February):1–2.

Cárdenas, Lázaro. 1978[1937]. Discurso del presidente de la républica al pueblo oaxaqueño. Oaxaca, Oax., 26 de Marzo de 1937. In *Palabras y documentos publicos de Lázaro Cárdenas 1928–1970: Mensajes, discursos, declaraciones, entrevistas y otros documentos 1928–1940*, ed. E. V. Gómez and C. Valcarce. Mexico City: Siglo Veintiuno Editores S.A.

Castillo Roman, Adriana. 1996. Se atienden todas las demandas por ilicitos que dañan los recursos. *El Nacional*, July 16, 1996.

CEMASREN. 1999. *Programa de manejo forestal para el aprovechamiento de recursos forestales maderables en la comunidad de Pueblos Mancomunados de Lachatao, Amatlán, Yavesía y Anexos, de los mismos municipios*. Oaxaca, Mexico: Distrito de Ixtlán, Estado de Oaxaca.

Challenger, Antony. 1998. *Utilización y conservación de los ecosistemas terrestres de México: Pasado, presente, y futuro*. Mexico City: CONABIO.

Chambille, Karel. 1983. *Atenquique: los bosques del sur de Jalisco, Los Bosques de México: Coleccion grandes problemas nacionales*. Mexico: Instituto de Investigaciones Economicas.

Chance, John K., and William B. Taylor. 1985. Cofradías and cargos: An historical perspective on the Mesoamerican civil-religious hierarchy. *American Ethnologist* 12 (1):1–26.

Chance, John K. 1989. *Conquest of the Sierra: Spaniards and Indians in colonial Oaxaca*. Norman, OK: University of Oklahoma Press.

Chapela, Francisco, and Yolanda Lara. 1993. *Impacto de la politica forestal sobre el valor de los bosques; el caso de la Sierra Norte de Oaxaca, México*. Oaxaca: Estudios Rurales Y Asesoria.

Chapela, Francisco, and Yolanda Lara. 1995. El papel de las comunidades campesinas en la conservación de los bosques. In *Cuadernos para una silvicultura sostenible*: Serie Sociedad y Politica No 1. Mexico City: Concejo Civil Mexicano Para La Silvicultura Sostenible.

Chapela, Gonzalo. 1997. El cambio liberal del sector forestal en México: Un analisis comparativo Canada-México-Estados Unidos. In *Semillas para el cambio en el campo. Medio ambiente, mercados y organización campesina*, ed. L. Pare, D. B. Bray, et al. Mexico City: Instituto de Investigaciónes Sociales, UNAM-IIS, La Sociedad de Solidaridad Social "Sansekan Tinemi" y Saldebas, Servicios de Apoyo Local al Desarrollo de Base en México, A.C.

Chassen, Francie R., and Hector G. Martinez. 1990. El desarrollo economico de Oaxaca a finales del Porfiriato. In *Lecturas historicas del Estado de Oaxaca, 1877–1930*, ed. M. A. R. Frizzi. Oaxaca: INAH, Gobierno de Oaxaca.

Chirinos, Luis, and Antonio Azuela. 1993. *La Urbanización popular y el orden jurídico en América Latina*. 1st. ed. Mexico City: Universidad Nacional Autónoma de México, Coordinación de Humanidades.

Chu, Julie. 2009. Card me when I'm dead: Identification papers and the pursuit of the good afterlife in China. In *Between life and death: Governing populations in the era of human rights*, ed. Sabine Berking and Magdalene Zolkos. Frankfurt: Peter Lang.

Presidente Municipal de Comaltepec. 1942. *En que se remite el programa en ocasion del día del árbol*. AGEO: Asuntos Agrarios, Problemas Por Bosques, Legajo 900, Expediente 16.

Compañía Forestal de Oaxaca. 1962. *Se rinde informe por guarda bosques*. AGEO: Asuntos Agrarios, Problemas Por Bosques, Legajo 897 Expediente 2.

Connolly, Priscilla. 1997. *El contratista de Don Porfirio: Obras publicas, deuda y desarrollo desigual*. Mexico: El Colegio de Michoacán, Universidad Autonoma Metropolitana Azcapotzalco, Fondo de Cultura Económica.

Contreras Arias, Alfonso. 1950a. Breves consideraciones acerca de la intervención del hombre en la modificación de las condiciones climáticas y de habitación de la Ciudad de México. *México Forestal* 28 (11–12: November–December): 85–89.

Contreras Arias, Alfonso. 1950b. Breves consideraciones acerca de la intervención del hombre en la modificación de las condiciones climáticas y de habitación de la Ciudad de México. *México Forestal* 28 (7–10: July–October):66–71.

Conzatti, Cassiano. 1914. La repoblación arborea del valle de Oaxaca. *Boletín de la Estación Agrícola Experimental de Oaxaca Numero* 1:1–13.

Coronil, Fernando. 1997. *The magical state, nature, money, and modernity in Venezuela*. Chicago: University of Chicago Press.

Craib, Raymond B. 2004. *Cartographic Mexico: A history of state fixations and fugitive landscapes*. Durham, NC: Duke University Press.

Daston, Lorraine, and Peter Galison. 1992. The image of objectivity. *Representations* 40:81–128.

de la Fuente, Julio. 1949. *Yalalag, una villa zapoteca serrana*. Mexico City: Museo Nacional de Antropología.

de la Vega, Ricardo. 1933. El fuego y la expansión agraria hacia el bosque como principales motivos de la deforestación del territorio patrio. *México Forestal* 11 (11–12):205–209.

de Lioucourt, F. 1898. De l'amenagement des sapinières. *Bul. Soc. For. Franche-Compte Belf*ort (juillet), 396–409.

de Vos, Jan. 1996. *Oro verde: La conquista de la Selva Lacandona por los madereros Tabasqueños 1822–1949*. Mexico: Fondo de Cultura Económica.

Dennis, Philip Adams. 1987. *Intervillage conflict in Oaxaca*. New Brunswick, NJ: Rutgers University Press.

Departamento de Bosques. 1912. *Operaciones forestales o dasocracia. vol. numero 5, Cartilla forestal, o resumen de la enseñanza que se da a los alumnos de la escuela nacional forestal: Escuela Nacional Forestal*. Secretaría de Fomento.

Desrosières, Alain. 1991. How to make things which hold together: social science, statistics, and the state. In *Discourses on society: The shaping of the social science disciplines*, ed. P. Wagner, B. Wittrock, and R. Whitley. Dordrecht: Kluwer.

DeWalt, Billy R., and Fernando Guadarrama Olivera. 2000. *Evaluación a medio camino del Proyecto de Conservación y Manejo Sustentable de Recursos Forestales en México*. Oaxaca: PROCYMAF Oaxaca.

Díaz Jiménez, Rodolfo. 2000. Consumo de leña en el sector residencial de Mexico: Evolución histórica de emisiones de CO2. M.S., Division de Estudios de Posgrado, Facultad de Ingeniería, Universidad Nacional Autónoma de México, Mexico.

Dirección Forestal y de Caza y Pesca. 1930. *Instrucciónes para la campaña contra incendios de montes*. AGEO: Asuntos Agrarios, Serie V Problemas por Bosques.

Dirección Forestal y de Caza y Pesca. 1932. *Circular relativa a la campaña contra incendios de monte girada a todos los presidentes municipales del estado*. AGEO: Asuntos Agrarios. Serie V, Problemas Por Bosques, Legajo 893, Expediente 11.

Donovan, D. G. 1981. *Fuelwood: How much do we need?* Hanover, NH: Institute of Current World Affairs.

Douglas, Mary. 1986. *How institutions think*. Syracuse, NY: Syracuse University Press.

Dove, M. 1983. Theories of swidden agriculture and the political economy of ignorance. *Agroforestry Systems* 1 (2):85–99.

Dove, M. R. 1994. The transition from native forest rubbers to *Hevea Brasiliensis* (Euphorbiaceae) among tribal smallholders in Borneo. *Economic Botany* 48 (4):382–396.

Dove, Michael R. 2006. Indigenous people and environmental politics. *Annual Review of Anthropology* 35 (1):191–208.

Dumit, J. 2003. Ways of seeing brains as expert images. In *Picturing personhood: Brain scans and biomedical identity*. Princeton, NJ: Princeton University Press.

Enríquez Quintana, Manuel. 1976. Las empresas ejidales forestales. *Revista de México Agrario* 9 (2):71–95.

Escárpita Herrera, Abraham. 1959. Especificaciones y adopción de una clave nacional para la fotointerpretación: Recomendaciones para su elaboración. *El Mensajero Forestal* 18 (181):14–31.

Escárpita Herrera, Jaime. 1980. Aspectos forestales de la porción oaxaqueña de la cuenca del Papaloapan. Paper read at *Junta de Trabajo sobre Desarrollo de la Cuenca del Papaloapan*, 12/29/80, at Tuxtepec, Oaxaca.

Escobar, Arturo. 1991. Anthropology and the development encounter: The making and marketing of development anthropology. *American Ethnologist* 18 (4):658–682.

Escobar, Arturo. 1995. *Encountering development: The making and unmaking of the Third World*. Princeton, NJ: Princeton University Press.

Espín Díaz, Jaime L. 1986. *Tierra fría, tierra de conflictos en Michoacan*. Zamora, Michoacán, México: Colegio de Michoacán.

Ezrahi, Yaron. 1990. *The descent of Icarus: Science and the transformation of contemporary democracy*. Vol. 12. Cambridge, MA: Harvard University Press.

Fairhead, James, and Melissa Leach. 2000. *Misreading the African landscape: Society and ecology in a forest-savanna mosaic*. Cambridge: Cambridge University Press.

FAO and CIFOR. 2005. Forests and floods. Drowning in fiction or thriving on facts? In *RAPP Publication 2005/03 Forest Perspectives* 2. Bogor, Indonesia.

FAO. 2001. FAOSTAT database. Rome, Italy: FAO.

FAPATUX. 1977. *Estudio dasonómico de las comunidades de San Juan Bautista Atepec y Miguel Aloapam*. Oaxaca, Mexico.

FAPATUX. 1956. Memoria descriptiva de la unidad industrial forestal para las Fábricas de Papel Tuxtepec, S.A. de C.V. en los estados de Oaxaca y Veracruz. Tuxtepec.

Ferguson, J. 1994. *The anti-politics machine. "Development" depoliticization and bureaucratic power in Lesotho*. Minneapolis: University of Minnesota Press.

Fierros, Aurelio M., and Saul Monreal Rangel. 2000. *La profesión forestal en México* (1909–2000) (*Crónica de una profesión desconocida*). Mexico City.

Fischer, Michael M. J. 2007. Four genealogies for a recombinant anthropology of science and technology. *Cultural Anthropology* 22:539–615.

Forsyth, Tim, and Andrew Walker. 2008. Forest guardians, forest destroyers. Seattle: University of Washington Press.

Foucault, Michel. 1979. Panopticism. In *Discipline and punish*. New York: Vintage.

Foucault, Michel. 1991. Truth and power. In *Power/knowledge: Selected interviews and other writings*. ed. and trans. Colin Gordon. Brighton, Sussex: Harvester Press.

Franklin, Sarah. 2007. *Dolly mixtures: The remaking of genealogy (A John Hope Franklin Center Book)*. Durham, NC: Duke University Press.

Friedlander, Judith. 1981. The secularization of the Cargo system: An example from postrevolutionary Central Mexico. *Latin American Research Review* 16 (2):132–143.

Fulé, Peter Z., and W. W. Covington. 1999. Fire regime changes in La Michilia Biosphere Reserve, Durango, Mexico. *Conservation Biology* 13 (3):640–652.

Fulé, Peter Z., José Villanueva-Díaz, and Mauro Ramos-Gómez. 2005. Fire regime in a conservation reserve in Chihuahua, Mexico. *Canadian Journal of Forest Research* 35:320–330.

García, Angel. 1996. *Oaxaca: Distritos, municipios, localidades y habitantes*. Oaxaca: Angel García García y Colaboradores.

García Toledo, A. 1930. *Que para al corte de leña y madera a que se refiere, debe cumplirse previamente con lo dispuesto en articulo de Ley Forestal*. AGEO: Asuntos Agrarios. Serie V, Problemas Por Bosques, Legajo 900, Expediente 12.

García Pérez, Pedro Vidal. 2000. *La region de la Sierra Juarez: Las propiedades comunales y el desarrollo sustentable*. Oaxaca, Mexico: PROCYMAF, SEMARNAP, WWF.

Garner, Paul. 1988. *La revolución en provincia: Soberanía estatal y caudillismo en las montañas de Oaxaca (1910–1920)*. Oaxaca, Mexico: Fondo de Cultura Economica.

Garner, Paul. 1990. Federalismo y caudillismo en la revolución Mexicana: Génesis del movimiento de soberanía en Oaxaca: 1915–1920. In *Lecturas historicas del estado de Oaxaca. 1877–1930*, ed. Maria de los Angles Romero Frizzi. Oaxaca: INAH, Gobierno de Oaxaca.

Gibson, Charles. 1964. *The Aztecs under Spanish rule: A history of the indians of the valley of Mexico 1519–1810*. Stanford: Stanford University Press.

Gieryn, Thomas F. 1995. The boundaries of science. In *Handbook of science and technology studies*, ed. Sheila Jasanoff et al. Thousand Oaks, CA: Sage.

Gijsbers, Wim. 2000. Teatro Crisol. *Noticias Ecología* 4/21 (00):8.

Glockner, Julio. 1999. Pedidores de lluvia del altiplano central Mexicano. *Scripta Ethnologica* 21:133–140.

Gobierno de Oaxaca. 2003. *Cámara nacional de la industria forestal*. http://www.oaxaca.gob.mx/sedic/forestal/spanish/dir-for.html.

Goldammer, J. G. 1993. Historical bioegeography of fire: tropical and subtropical. In *Fire in the environment: The ecological, atmospheric, and climatic importance of vegetation fires*, ed. P. J. Crutzen and J. G. Goldammer. Chichester, England: John Wiley and Sons.

Goldammer, J. G., and B. Seibert. 1990. The impact of droughts and forest fires on tropical lowland rain forest of East Kalimantan. In *Fire in the tropical Biota: Ecosystem processes and global challenges*, ed. J. G. Goldammer. New York: Springer-Verlag.

Goldman, Michael. 2001. The birth of a discipline: Producing authoritative green knowledge, World Bank-style. *Ethnography* 2 (2):191–217.

Goloubinoff, Marina, Esther Katz, and Annamaria Lammel, eds. 1997. *Antropología del clima en el mundo Hispanoamericano, tomo I*. Colección Biblioteca Abya-Ayala. Quito, Ecuador: Ediciones Abya-Yala.

Gómez Cárdenas, Martín, Porfirio López López, Juan F. Castellanos Bolaños, and Aurelio Manuel Fierros González. 1999. *Influencia de los aprovechamientos maderables en la diversidad vegetal en la Sierra Norte de Oaxaca*. Oaxaca, Mexico: INIFAP, Campo Experimental Valles Centrales.

Gómez, Marte R. 1930. *Memorandum to Dirección Forestal y de Caza y Pesca: Disposiciones reglamentarias para efectuar las "quemas de limpia."* AGEO: Asuntos Agrarios, Problemas Por Bosques, Serie V, Legajo 892, Expediente 22.

Gómez Mena, Carolina. 2003. *Solicitó a la Sagarpa que "decrete algún tipo de figura jurídica": Plantea la Conafor quitar subsidios a agricultores que quemen terrenos. La Jornada* 3 (27).

González, Roberto J. 2001. *Zapotec science*. Austin, TX: University of Texas Press.

González-Espinosa, Mario, Pedro F. Quintana-Ascencio, Neptali Ramírez-Marcial, and Patricia Gaytan-Guzmán. 1991. Secondary succession in disturbed Pinus-Quercus forests in the highlands of Chiapas, Mexico. *Journal of Vegetation Science* 2:351–360.

Graeber, David. 2006. Beyond power/knowledge: An exploration of the relationship of power, ignorance, and stupidity. Malinowski Memorial Lecture, London School of Economics and Political Science.

Grindle, Merilee Serrill. 1977. *Bureaucrats, politicians, and peasants in Mexico: A case study in public policy*. Berkeley: University of California Press.

Grove, Richard H. 1995. *Green imperialism*. New York: Cambridge University Press.

Grupo de Estudios Ambientales. 2001. Film. *Voces del monte: Experiencias comunitarias para el manejo sustentable de los bosques en Oaxaca*. Mexico City: SEMARNAP/PROCYMAF.

Guardino, Peter. 2000. "Me ha cabido en la fatalidad" Gobierno indígena y gobierno republicano en los pueblos indígenas: Oaxaca, 1750–1850. *Desacatos* 5:119–130.

Gupta, Akhil. 1995. Blurred boundaries: The discourse of corruption, the culture of politics, and the imagined state. *American Ethnologist* 22 (2):375–402.

Gupta, Akhil. 1998. *Postcolonial developments*. London: Duke University Press.

Gupta, Akhil, and Aradhana Sharma. 2006. Globalization and postcolonial states. *Current Anthropology* 47 (2): 227–307.

Gutiérrez, José L. 1930. *Para prevenir y combatir los incendios de montes en la República, esta Dirección ha estado organizando el mayor numero posible de Corporaciones de Defensa Contra Incendios de Montes*. AGEO: Asuntos Agrarios. Serie V, Problemas Por Bosques, Legajo 892, Expediente 16.

Gutiérrez, Velasco, and Faustino Froylan. 1986. *La explotación forestal y sus repercuciones en una comunidad Zapoteca: Atepec, distrito de Ixtlán*. Oaxaca: INI.

Haas, Peter M. 1990. *Saving the Mediterranean: The politics of international environmental cooperation, the political economy of international change*. New York: Columbia University Press.

Haenn, Nora. 2002. Nature regimes in Southern Mexico: A history of power and environment. *Ethnology* 41 (1):1–26.

Haenn, Nora. 2005. *Fields of power, forests of discontent: Culture, conservation, and the state in Mexico*. Tucson: University of Arizona Press.

Hajer, Maarten A. 1995. Discourse analysis. In *The politics of environmental discourse: Ecological modernization and the policy process*, 42–72. Oxford: Oxford University Press.

Hamnett, Bryan R. 1971. Dye production, food supply, and the laboring population of Oaxaca, 1750–1820. *Hispanic American Historical Review* 51:51–78.

Haraway, Donna. 1991. Situated knowledges: The science question in feminism and the privilege of partial perspective. In *Simians, cyborgs, and women*. New York: Routledge.

Haraway, Donna. 2008. *When species meet*. Minneapolis: University of Minnesota Press.

Hayden, Cori. 2003. *When nature goes public: The making and unmaking of bioprospecting in Mexico*. Princeton, NJ: Princeton University Press.

Helmreich, Stefan. 2008. *Alien ocean: Anthropological voyages in microbial seas*. Berkeley: University of California Press.

Hémond, Aline, and Marina Goloubinoff. 1997. El "Via Crucis" del Agua: Clima, Calendario Agricola y Religioso Entre los Nahuas de Guerrero (México). In *Antropología del clima en el mundo Hispanoamericano*, vol. 1. Ecuador: Ediciones Abya-Yala.

Herzfeld, Michael. 1992. *The social production of indifference: Exploring the symbolic roots of Western bureaucracy*. New York: Berg.

Herzfeld, Michael. 1997. *Cultural intimacy: Social poetics in the nation-state*. New York: Routledge.

Herzfeld, Michael. 2005. Political optics and the occlusion of intimate knowledge. *American Anthropologist* 107 (3):369–376.

Heyerdahl, Emily K., and Ernesto Alvarado. 2003. Influence of climate and land use on historical surface fires in pine-oak forests, Sierra Madre Occidental, Mexico. In *Fire and climatic change in temperate ecosystems of the western Americas*, ed. T. T. Veblen, W. L. Baker, G. Montenegro, and T. W. Swetnam. New York: Springer-Verlag.

Hilgartner, S. 2000. *Science on stage: Expert advice as public drama.* Stanford, CA: Stanford University Press.

Hinojosa Ortíz, Manuel. 1958. *Los bosques de Mexico: Relato de un despilfarro y una injusticia.* Mexico City: Instituto Mexicano de Investigaciones Económicas.

Ho, Karen. 2005. Situating global capitalisms: A view from Wall Street investment banks. *Cultural Anthropology* 20 (1):68–96.

Hobart, Mark, ed. 1993. *An anthropological critique of development: The growth of ignorance.* London: Routledge.

Huguet, L. 1953. Unidades industriales de explotación forestal: A system of organized forest exploitation practised in Mexico. *Unasylva* 7 (2):50–54.

Hull, Matthew S. 2003. The file: Agency, authority, and autography in an Islamabad bureaucracy. *Language & Communication* 23 (3–4):287–314.

Huntingdon, Samuel P. [1968] 2002. Modernization and corruption. In *Political corruption: Concepts and contexts*, ed. A. J. Heidenheimer and M. Johnston. New Brunswick, NJ: Transaction.

Ibarra, Isaac M. 1975. *Memorias del general Isaac M. Ibarra.* Mexico City: Autobiografia.

INEGI. 1980a. *Carta de precipitación total anual.* Mexico City.

INEGI. 1980b. *Carta de temperaturas medias anuales.* Mexico City.

INEGI. 1990a. Climas. In *Atlas nacional de México.* Mexico: UNAM.

INEGI. 1990b. Hipsometría y Batimetría. In *Atlas nacional de Mexico.* Mexico: UNAM.

INEGI. 1990c. Vegetación. In *Atlas nacional de Mexico.* Mexico: UNAM.

INEGI.1995. ORTOFOTO DIGITAL: E14D38E. Fotografías aéreas escala 1:75,000.

INEGI. 2003. XII Censo General de Población y Vivienda 2000, Sistema municipal de bases de datos: SIMBAD. http://www.inegi.gob.mx/difusion/espanol/fiest.html.

InfoLatina. 1998. Plantar 500 mil arboles en Toluca, meta en 1998. *El Diario de Toluca*, July 10, 1998.

Jasanoff, S. 1998. The eye of everyman: Witnessing DNA in the Simpson Trial. *Social Studies of Science* 28 (5–6):713–740.

Jasanoff, S. 2004a. Ordering knowledge, ordering society. In *States of knowledge: The co-production of knowledge and social order*. London: Routledge.

Jasanoff, S. 2004b. *States of knowledge: The co-production of knowledge and social order*. London: Routledge.

Jasanoff, S. 2004c. Heaven and Earth: The politics of environmental images. In *Earthly politics: Local and global in environmental governance*. Cambridge: MIT Press.

Jasanoff, S. 2005. *Designs on nature*. Princeton, NJ: Princeton University Press.

Jayo Ceniceros, Alejandro. 1980. Plantaciones para uso domestico. In Conference Proceedings, *Algunas experiencias personales sobre reforestación*. Oaxtepec, Morelos, Mexico.

Joseph, Gilbert M., and Daniel Nugent. 1994a. Popular culture and state formation in Revolutionary Mexico. In *Everyday forms of state formation: Revolution and the negotiation of rule in modern Mexico*, ed. G. M. Joseph and D. Nugent. Durham, NC: Duke University Press.

Joseph, Gilbert M., and Daniel Nugent, eds. 1994b. *Everyday forms of state formation: Revolution and the negotiation of rule in modern Mexico*. Durham, NC: Duke University Press.

Junta Central de Bosques y Arboledas. 1911. Reforestación. Vol. Numero 3, Cartilla forestal, o resumen de la enseñanza que se da a los alumnos aspirantes al cargo de guardia forestal. Mexico City: Secretaria de Fomento.

Junta Central de Bosques y Arboledas. 1912. Viveros de Arboles. Vol. Numero 1, Cartilla forestal, o resumen de la enseñanza que se da a los alumnos de la escuela nacional forestal: Escuela Nacional Forestal. Mexico City: Secretaría de Fomento.

Kitzberger, Thomas, Peter M. Brown, Emily K. Heyerdahl, Thomas W. Swetnam, and Thomas T. Veblen. 2007. Contingent Pacific-Atlantic Ocean influences on multicentury wildfire synchrony over western North America. *Proceedings of the National Academy of Sciences of the United States of America* 104 (25):543–548.

Klepeis, Peter, and B. L. Turner, II. 2001. Integrated land history and global change science: The example of the Southern Yucatan Peninsular Region. *Land Use Policy* 18:27–39.

Klooster, D. 2000. Community forestry and tree theft in Mexico: Resistance or complicity in conservation? *Development and Change* 31 (1):281–305.

Klooster, D. 2003a. Campesinos and Mexican forest policy during the twentieth century. *Latin American Research Review* 38 (2):94–126.

Klooster, D. 2003b. Social nature: Theory, practice, and politics. *Annals of the Association of American Geographers. Association of American Geographers* 93 (4):938–941.

Kull, Christian A. 2002. Madagascar's burning issue: The persistent conflict over fire. *Environment* 44 (3):8–19.

Kull, Christian A. 2005. Historical landscape repeat photography as a tool for land use change research. Norsk Geografisk Tidsskrift (Norwegian *Journal of Geography)* 59 (4):253–268.

Lartigue, Francois. 1983. *Indios y bosques: politicas forestales y comunales en la Sierra Tarahumara, Ediciones de la Casa Chata.* Centro de Investigaciones y Estudios Superiores en Antropología Social.

Latour, Bruno. 1987. *Science in action: How to follow scientists and engineers through society.* Cambridge, MA: Harvard University Press.

Latour, Bruno. 1990. Drawing things together. In *Representation in scientific practice*, ed. M. Lynch and S. Woolgar. Cambridge, MA: MIT Press.

Latour, Bruno. 1993. *We have never been modern.* Cambridge, MA: Harvard University Press.

Latour, Bruno. 2004. Why has critique run out of steam?: From matters of fact to matters of concern. *Critical Enquiry* 30 (Winter):225–248.

Leon, Marcos F. 1983. Pediran a MMH que se anule la concesión a FAPATUX y a la Cía. Forestal de Oaxaca. *Panorama*, February 1983.

Lerer, L. B., and T. Scudder. 1999. Health impacts of large dams. *Environmental Impact Assessment Review* 19 (2):113–123.

Li, Tania. 1999. Compromising power: Development, culture, and rule in Indonesia. *Cultural Anthropology* 14 (3):295–322.

Li, Tania. 2000. Articulating indigenous identity in Indonesia: Resource politics and the tribal slot. *Comparative Studies in Society and History* 42 (1): 149–179.

Li, Tania. 2006. *The will to improve: Governmentality, development, and the practice of politics.* Durham, NC: Duke University Press.

Lomnitz, Claudio. 1995. Ritual, rumour, and corruption in the constitution of the polity in modern Mexico. *Journal of Latin American Anthropology* 1 (1):20–47.

Lomnitz, Larissa. 1982. Horizontal and vertical relations and the social structure of urban Mexico. *Latin American Research Review* 17 (2):51–74.

Lomnitz, Larissa, Leticia Mayer, and Martha W. Rees. 1983. Recruiting technical elites: Mexico's veterinarians. *Human Organization* 42 (1):2–29.

Lopez, Vicente G. 1930. *Solicita se le ministre la Ley Forestal por carecer de ella.* AGEO: Asuntos Agrarios, Serie V, Problemas Por Bosques, Legajo 900, Expediente 13.

López Cortes, Francisco. 1930. *Que ya he ordenado a la autoridad municipal de Teococuilco exija el cumplimiento de las disposiciones sobre quemas de limpia.* AGEO: Asuntos Agrarios, Serie V, Problemas Por Bosques, Legajo 900, Expediente14.

Lorence, David H., and Abisai García Mendoza. 1989. Oaxaca, Mexico. In *Floristic inventory of tropical countries*, ed. D. G. Campbell and D. H. Hamond. New York: New York Botanical Garden.

MacKenzie, Donald. 2008. *Material markets: How economic agents are constructed.* Clarendon Lectures in Management Series. Oxford: Oxford University Press.

Mairesse, Charles. 1880. Report on the silver mines of Cinco Señores San Geronimo situated in the district of Ixtlán, or Villa-Juárez in the state of Oaxaca. [Mexico City: Filomeno Mata]. Vol. Numeros 5:6.

Mallon, Florencia E. 1994. Reflections on the ruins: Everyday forms of state formation in nineteenth century Mexico. In *Everyday forms of state formation: Revolution and the negotiation of rule in modern Mexico*, ed. G. M. Joseph and D. Nugent. Durham, NC: Duke University Press.

Mamdani, Mahmood. 1996. *Citizen and subject: Contemporary Africa and the legacy of late colonialism.* Princeton, NJ: Princeton University Press.

Mares, José. 1932. *Se remite plan general para la campaña contra incendios de montes.* AGEO: Asuntos Agrarios, Serie V, Problemas Por Bosques, Legajo 895, Expediente 21.

Martínez Luna, Jaime. 1977. Aqui el que manda es el pueblo. *Cuadernos Agrarios* Ano 2 (5, September):98–113.

Martínez Luna, Jaime. 2004. Comunalidad y desarrollo. In Diálogos en la Acción, Segunda Etapa, 339–354.

Masera, Omar, Jaime Navia, Teresita Arias Chalico, and Enrique Riegelhaupt. 1997. *Patrones de consumo de leña en tres micro-regiones de Mexico: Síntesis de resultados.* Mexico City: Proyecto FAO/Mexico "Dendroenergia para el desarrollo rural."

Mathews, Andrew Salvador. 2003. Suppressing fire and memory: Environmental degradation and political restoration in the Sierra Juárez of Oaxaca 1887–2001. *Environmental History* 8 (1):77–108.

Mathews, Andrew Salvador. 2004. Ph.D. Thesis. Forestry culture: Knowledge, institutions, and power in Mexican forestry, 1926–2001, Forestry and Environmental Studies, Anthropology Department, Yale University, New Haven, CT.

Mathews, Andrew Salvador. 2005. Power/knowledge, power/ignorance: Forest fires and the state in Mexico. *Human Ecology* 33 (6):795–820.

Mathews, Andrew Salvador. 2009. Unlikely alliances: Encounters between state science, nature spirits, and indigenous industrial forestry in Mexico, 1926–2008. *Current Anthropology* 50 (1):75–101.

Mecinas, Cutberto, and Jaime Martínez Luna. 1981. *Los problemas de la organización: El caso de la empresa comunal de los Pueblos Mancomunados de Lachatao, Amatlan, Yavesía y Anexos.* Cuadernos de Trabajo Num. 3. Oaxaca: Instituto Tecnologico de Oaxaca, Centro Regional de Graduados y Investigación Tecnológica.

McNamara, Patrick J. 2007. *Sons of the Sierra: Juárez, Díaz, and the people of Ixtlán, 1855–1920.* Oaxaca: University of North Carolina Press.

Meixueiro, Raymundo. 1954. *Explotación de Madera por la Compañia Aserradera de Raymundo Meixueiro*. AGEO: Asuntos Agrarios, Serie V, Problemas Por Bosques, Legajo 895, Expediente 17.

Mejía Fernández, Lázaro. 1988. Masters' Thesis. La política forestal en el desarrollo de la administración pública forestal. Tesis de Ingeniero. División de Ciencias Forestales, Universidad Autónoma de Chapingo, Mexico.

Mendoza Medina, Roberto. 1976. La política forestal en el sector ejidal y comunal. *Revista del México Agrario* 9 (2):35–69.

Merino, Leticia. 1997. *El manejo forestal comunitario y sus perspectivas de sustentabilidad*. Cuernavaca: UNAM.

Mitchell, Timothy. 2002. *Rule of experts: Egypt, techno-politics, and modernity*. Berkeley: University of California Press.

Moore, Donald S. 1998. Clear waters and muddied histories: Environmental history and the politics of community in Zimbabwe's eastern highlands. *Journal of Southern African Studies* 24 (2):377–403.

Moore, Donald S. 2005. *Suffering for territory: Race, place, and power in Zimbabwe*. Durham, NC: Duke University Press.

Morris, Stephen D. 1991. *Corruption and politics in contemporary Mexico*. Tuscaloosa: University of Alabama Press.

Mosse, David. 2005. *Cultivating development: An ethnography of aid policy and practice*. London; Ann Arbor, MI: Pluto Press.

Muro, Ricardo del. 1983. Conceden el amparo a los comuneros de Oaxaca contra un decreto presidenciál. Uno Mas Uno, February 27, 1983.

Musalem López, Francisco Javier. 1972. Breve analisis sobre la silvicultura y manejo de bosques de coniferas en México. *México y sus bosques* (November–December):21–32.

Musalem López, Francisco Javier. 1979. *Las bases y primeras acciones del programa nacional de mejoramiento silvicola en bosques de coníferas*. Morelia, Michoacán: PROFORMICH.

Musset, Alain. 1991. *De l'eau vive á l'eau morte: 1492–1992: Enjeux techniques et culturels dans la vallée de Mexico* (XIVe-XIXe siécles). Editions recherche sur les civilisations. Paris.

Nader, Laura. 1990. *Harmony ideology, justice, and control in a Zapotec mountain village*. Stanford, CA: Stanford University Press.

Negreros, P., and L. K. Snook. 1984. Análisis del efecto de la intensidad de corta sobre la regeneración de pinos en un bosque mezclada de pino encino. *Ciencia Forestal* 47:48–61.

Nelson, A. R. 2005. *Cold War ecology: Forests, farms, and people in the East German landscape, 1945–1989*. Yale Agrarian Studies Series. New Haven, CT: Yale University Press.

Nigh, Ronald Byron. 1975. Ph.D. Thesis. Evolutionary ecology of Maya agriculture in Highland Chiapas, Mexico. Anthropology Department, Stanford University, Stanford, CA.

Nuijten, Monique. 2003. *Power, community, and the state: The political anthropology of organisation in Mexico*. London: Pluto Press.

Nuijten, Monique, and Gerhard Anders. 2007. Introduction: Anthropology of corruption and the secret of law. In *Corruption and the secret of law*. London: Ashgate.

Nye, J. S. [1967] 2002. Corruption and political development: A cost-benefit analysis. In *Political corruption: Concepts and contexts*, ed. A. J. Heidenheimer and M. Johnston. New Brunswick, London: Transaction.

Oliver, Chadwick Dearing, and Bruce C. Larson. 1990. *Forest stand dynamics*. New York: McGraw-Hill.

Ortega Pizarro, Fernando, and Guillermo Correa. 1983. Ya de salida, López Portillo regaló bosques de Oaxaca a Pandal Graff: 200,000 indígenas condenados a la esclavitud. *Proceso*, February 7, 1983.

Ostrom, Elinor. 1991. *Governing the commons: The evolution of the institutions for collective action*. New York: Cambridge University Press.

Pauli, Daniel. 2001. Systematic distortions in world fisheries catch trends. *Nature* 414 (29):534–536.

Peña Manterola, Gonzalo. 1967. Fotografías del Recuerdo: Escuela Nacional Forestal de Coyoacán, D.F. Años de 1917, 1918 y 1919. Mexico Forestal XLII (1) (January–February) (Segunda Epoca):69–72.

Pérez García, Rosendo. 1996 [1956]a. *La Sierra Juarez*, 2nd ed., vol. 1: *Coleccion historia*. Oaxaca, Mexico: Instituto Oaxaqueño De Las Culturas.

Pérez García, Rosendo. 1996 [1956]b. *La Sierra Juarez*, 2nd ed., vol. 2: *Coleccion historia*. Oaxaca, Mexico: Instituto Oaxaqueño De Las Culturas.

Pinch, Trevor J., and Wiebe E. Bijker. 2002. The social construction of facts and artifacts: Or how the sociology of science and the sociology of technology might benefit each other. In *The social construction of technological systems*, eds. Wiebe Bijker, T. Pinch and T. Hughes. Cambridge: MIT Press.

Plancarte Barrera, Antonio. 2000. Los cambios cuantitativos más relevantes registrados en materia de recursos naturales entre 1995 y 1999. (Numeralia). Paper read at Foro forestal Oaxaca 2000, Hotel Fortin Plaza, Oaxaca, OAX.

Power, M. 1997. Audit knowledge and the construction of auditees. In *The audit society: Rituals of verification*. Oxford: Oxford University Press.

Presidencia de la República. 1958a. Decreto que declara de utilidad publica la constitución de una unidad industrial de explotación forestal en favor de la empresa denominada Fábricas de Papel Tuxtepec, S.A. de C.V. In Diario Oficial. Mexico: Código Forestal.

Presidencia de la República. 1958b. Código Forestal: Decreto que establece una Unidad Industrial de Explotación Forestal a favor de la Compañia Forestal de Oaxaca S. de R.L., en predios boscosos enclavados en los Distritos de Zimatlán, Sola de Vega, Miahuatlán y Yautepec, en el Estado de Oaxaca. *Diario Oficial* 10/15/58, 736–753.

Pyne, Steven J. 1993. Keeper of the flame: A survey of anthropogenic fire. In *Fire in the environment: Its ecological, climatic, and atmospheric chemical importance*, ed. P. J. Crutzen and J. G. Goldammer. Chichester, England: John Wiley.

Pyne, Steven J. 1998. Forged in fire: History, land, and anthropogenic fire. In *Advances in historical ecology*, ed. W. Balée. New York: Columbia University Press.

Quevedo, Miguel Angel de. 1916. *Algunas consideraciones sobre nuestro problema agrario*. Mexico City: Imprenta Victoria.

Quevedo, Miguel Angel de. 1926a. Las polvaredas de los terrenos tequezquitozos del antiguo Lago de Texcoco y los procedimientos de enyerbe para remediarlas. *Mexico Forestal* 5 (5–6, May–June):39–52.

Quevedo, Miguel Angel de. 1926b. Los desastres de la deforestación en el valle y Ciudad de México. *México Forestal* 4 (7–8, July–August):67–82.

Quevedo, Miguel Angel de. 1928. Los incendios de nuestros bosques y la necesaria atención para prevenirlos. *México Forestal* 6 (11, November):213–223.

Quevedo, Miguel Angel de. 1933. La ruina forestal en los lomeríos y serranías del suroeste del distrito federal. *México Forestal* 11 (7–8, July–August): 127–129.

Quevedo, Miguel Angel de. 1935 [1910]. El origen de la cuestión forestal en México: Espacios libres y reservas forestales de las ciudades, su adaptación a jardines, parques y lugares de juego, aplicación a la Ciudad de México. (Conferencia Sustentada en la Exposición de Higiene en Septiembre de 1910). *México Forestal* 13 (11–12, November–December):105–116.

Quevedo, Miguel Angel de. 1938. Las labores del departamento de 1935 a 1938. *Boletín del Departamento Forestal y de Caza y Pesca* 3 (August):39–79.

Quevedo, Miguel Angel de. 1941. La iniciación de la campaña de protección forestal del territorio nacional y sus desarrollos sucesivos y tropiezos. *México Forestal* 19 (7–8, July–August):67–76.

Quintana-Ascencio, P. F., M. González-Espinosa, and Neptali Ramírez-Marcial. 1992. Acorn removal, seedling survivorship, and seedling growth of Quercus crispipilis in successional forests of the highlands of Chiapas, Mexico. *Bulletin of the Torrey Botanical Club* 119 (1):6–18.

Raffles, Hugh. 2002. *In Amazonia: A natural history*. Princeton, NJ: Princeton University Press.

Rajan, S. Ravi. 2006. *Modernizing nature: Forestry and imperial eco-development 1800–1950*. Oxford: Oxford University Press.

Ramos, Fernando. 2000. En Sierra Juarez, obtienen premios al merito ecologico 14 comunidades. El Imparcial, 7/15/00, 7B.

Real Academia Española. 2003. *Gremio* (22). http://www.rae.es/rae.html.

Richardson, David M. 1998. *Ecology and biogeography of Pinus*. Cambridge: Cambridge University Press.

Richardson, David M., Philip W. Rundel, Stephen T. Jackson, Robert O. Teskey, James Aronson, Andrzej Bytnerowicz, Michael J. Wingfield, and Bûerban Procheü. 2007. Human impacts in pine forests: Past, present, and future. *Annual Review of Ecology Evolution and Systematics* 38 (1):275–297.

Riles, Annelise. 1998. Infinity within the brackets. *American Ethnologist* 25 (3):378–398.

Rizzoli, Angelo, and Vittorio De Sica. 2003 [1952]. *Umberto D.* Criterion Collection.

Robbins, Paul. 1998. Paper forests: Imagining and deploying exogenous ecologies in arid India. *Geoforum* 29 (1):69–86.

Robbins, Paul. 2003. Beyond ground truth: GIS and the environmental knowledge of herders, professional foresters, and other traditional communities. *Human Ecology* 31 (2):233–253.

Robbins, Paul. 2000. The practical politics of knowing: State environmental knowledge and local political economy. *Economic Geography* 76 (2):126–144.

Rocheleau, Dianne. 1995. Maps, numbers, text, and context: Mixing methods in feminist political ecology. *Professional Geographer* 47 (4):458–466.

Rodríguez-Trejo, Dante Arturo. 2008. Fire regimes, fire ecology, and fire management in Mexico. *AMBIO: A Journal of the Human Environment* 37 (7):548–556.

Romero Frizzi, María de los Angeles. 1988. Epoca Colonial (1519–1785). In *Historia de la cuestion agraria Mexicana: Estado de Oaxaca, prehispanico–1924*, ed. L. Reina. Oaxaca, Mexico: Centro de Estudios Historicos del Agrarismo en México, Universidad Autonoma Benito Juárez de Oaxaca.

Rosales Salazar, Paulino H., et al. 1982. El Metodo de desarrollo silvicola: Una alternativa en la silvicultura y ordenación de bosques.

Rus, Jan. 1994. The "Comunidad Revolucionaria Institucional": The subversion of native government in highland Chiapas, 1936–1968. In *Everyday forms of state formation: Revolution and the negotiation of rule in modern Mexico*, ed. G. M. Joseph and D. Nugent. Durham, NC: Duke University Press.

Rzedowski, J. 1978a. Bosque de coníferas. In *Vegetación de México*. D.F., Mexico: Editorial Limusa.

Rzedowski, J. 1978b. *Vegetación de México*. D.F., Mexico: Editorial Limusa.

Saberwal, Vasant K. 1997. Bureaucratic agendas and conservation policy in Himachal Pradesh, 1865–1994. *Indian Economic and Social Review* 34 (4):466–498.

SAG. 1967. *Memorias de labores de la Secretaría de Agricultura y Ganadería*. Mexico City: Talleres Gráficos de la Nacion.

SARH. 1979. *Desarrollo forestal*. Mexico City: Subsecretaría Forestal y de la Fauna, Dirección General Para el Desarrollo Forestal.

SARH. 1994. *Manual de procedimientos para la autorización del aprovechamiento de recursos forestales maderables y criterios para la dictaminación de programas de manejo forestal*. Mexico City: SARH.

SARH and UCODEFO #6. 1993. *Programa de Manejo Forestal para la comunidad de Pueblos Mancomunados de Lachatao, Amatlan, Yavesía y Anexos.* Oaxaca, Mexico: SARH.

Schama, Simon. 1995. *Landscape and memory*. London: Vintage.

Scott, James C. 1969. The analysis of corruption in developing nations. *Comparative Studies in Society and History* 11 (3):315–341.

Scott, James C. 1985. *Weapons of the weak: Everyday forms of peasant resistance*. New Haven, CT: Yale University Press.

Scott, James C. 1990. *Domination and the arts of resistance: Hidden transcripts.* New Haven, CT: Yale University Press.

Scott, James C. 1998. *Seeing like a state: How certain schemes to improve the human condition have failed*. New Haven: Yale University Press.

Segura, G., and L. C. Snook. 1992. Stand dynamics and regeneration patterns of a Pinyon pine forest in East Central Mexico. *Forest Ecology and Management* 47:175–194.

SEMARNAP. 1997a. *Anuario estadistico de la producción forestal*. Mexico City: SEMARNAP.

SEMARNAP. 1997b. México: Ley Forestal. Mexico City: SEMARNAP.

SEMARNAP. 1997c. NORMA Oficial Mexicana NOM-015-SEMARNAP/SAGAR-1997 Que regula el uso del fuego en terrenos forestales y agropecuarios.

SEMARNAP. 1996. NORMA Oficial Mexicana NOM-012-RECNAT-1996, Que establece los procedimientos, criterios y especificaciones para realizar el aprovechamiento de leña para uso doméstico.

SEMARNAP. 2001. Estructura organica de SEMARNAP.

SEMARNAP. 2002. Acuerdo por el que se establecen las especificaciones, procedimientos, lineamientos tecnicos y de control para el aprovechamiento, transporte, almacenamiento y transformación que identifique el origen legal de las materias primas forestales.

Senado de la República Mexicana. 1998. Sesion Plenaria del Senado de le República Mexicana: Version Estenográfica. D.F.

Serrato, Gilberto A. 1931. Acción de los montes sobre las cuencas hidrográficas, manantiales y vasos de almacenamiento. *Mexico Forestal* 9 (4) (April):98–101.

Shapin, Steven, and Simon Schaffer. 1985. *Leviathan and the air-pump: Hobbes, Boyle, and the experimental life.* Including a translation by Simon Schaffer of Thomas Hobbes, *Dialogus physicus de natura aeris*. Princeton, NJ: Princeton University Press.

Sharma, Aradhana, and Akhil Gupta. 2006. *The anthropology of the state: A reader*. Malden, MA: Blackwell.

Simonian, Lane. 1995. *Defending the land of the jaguar*. Austin: University of Texas Press.

Sivaramakrishnan, K. 2000. State sciences and development histories: Encoding local forestry knowledge in Bengal. *Development and Change* 31 (1):61–89.

Smith, Benjamin. 2002. Defending "our beautiful freedom": State formation and local autonomy in Oaxaca, 1930–1940. *Estudios Mexicanos* 23 (1):125–153.

Smith, Benjamin. 2008. Inventing tradition at gunpoint: Culture, caciquismo, and state formation in the region Mixe, Oaxaca (1930–1959). *Bulletin of Latin American Research* 27 (2):215–234.

Smith, Carol. 1996. Myths, intellectuals, and race/class/gender distinctions in the formation of Latin American nations. *Journal of Latin American Anthropology* 2 (1):148–169.

Smith, David M. 1997. *The practice of silviculture: Applied forest ecology*. New York: Wiley.

Snook, L. K. 1986. Effects of Mexico's selective cutting system on pine regeneration and growth in a mixed pine-oak (Pinus-Quercus) forest. *USDA Forest Service General Technical Report* SE-46:27–31.

Snook, L. K. 1995. Community forestry, sustainability, and forest conservation in Mexico: Past experiences and the implications of recent policy changes. *Manuscript*: 41.

Sociedad Mexicana de Historia Natural. 1883. Dictamen sobre la repoblación vegetal del Valle de México. *La Naturaleza* V:245–251.

Star, Susan Leigh, and James R. Griesemer. 1989. Institutional ecology, "translations," and boundary objects: Amateurs and professionals in Berkeley's Museum of Vertebrate Zoology, 1907–1939. *Social Studies of Science* 19 (3):387–420.

Strathern, Marilyn. 2000. Introduction: New accountabilities; Afterword: Accountability and ethnography. In *Audit cultures. Anthropological studies in accountability, ethics, and the academy*. London: Routledge.

Szekely, Miguel E., and Sergio Madrid. 1990. La apropiación comunitaria de recursos naturales: Un caso de la Sierra de Juárez, Oaxaca. In *Recursos naturales, técnica y cultura: Estudio y experiencia para un desarrollo sustentable*, ed. Enrique Leff and J. Carabias. Mexico City: UNAM, Coordinación de Humanidades.

Tamayo, Jorge L. 1976. Una experiencia forestal industrial. *Revista del México Agrario Año* 9 (2):145–158.

Tamayo, Jorge L. 1982 [1954]. Origen, creación, presente y futuro de la Comisión del Papaloapan: Diez años después de la pavorosa inundación, Tuxtepec emerge vigorosa (1944–1954). In Obras VII. Realidades y Proyecciones de Oaxaca. Mexico: Centro de Investigación Cientifica Jorge L. Tamayo.

Tamayo, Jorge L., and Enrique Beltran, eds. 1977. *Recursos naturales de la cuenca del Papaloapan*. 2 vols. Mexico: El Instituto.

Thompson, Charis. 2004. Co-producing CITES and the African elephant. In *States of knowledge: The co-production of knowledge and social order*. London: Routledge.

TIASA. 1993. *Programa de manejo forestal, Ixtlán de Juarez, Oaxaca, 1993–2000. 4 vols.* Mexico: Texcoco.

Trabulse, Elias. 1992. *José Maria Velasco: Un paisaje de la ciencia en México.* Toluca: Instituto Mexiquense de Cultura.

Treviño Saldaña, Carlos. 1937. Proyecto de Ordenacion de los Bosques de Atlamaxac, Puebla. *Boletín del Departamento Forestal y de Caza y Pesca* (8):177–253.

Tsing, Anna. 2005. *Friction: An ethnography of global connection, cultural anthropology.* Princeton, NJ: Princeton University Press.

Tsing, Anna. 2000. Inside the economy of appearances. *Public Culture* 12 (1):115–144.

Turner, Mathew D., and Peter J. Taylor. 2003. Critical reflections on the use of remote sensing and GIS technologies in human ecological research. *Human Ecology* 31 (2):177–182.

Van der Sluijs, Jeroen, Simon Shackley, and Brian Wynne. 1998. Anchoring devices in science for policy. *Social Studies of Science* 28 (2):291–323.

Van Ufford, Philip Quarles. 1993. Knowledge and ignorance in the practices of development policy. In *An anthropological critique of development: The growth of ignorance*, ed. Mark Hobart. London: Routledge.

Varese, Stefano. 1983. *Proyectos étnicos y proyectos nacionales.* Mexico City: SEP.

Various. 1957. *Relacionado con la explotación forestal.* AGEO: Asuntos Agrarios.

Various. 1958. *Tala de bosques en diferentes partes del estado.* AGEO: Asuntos Agrarios, Serie V, Problemas Por Bosques, Legajo 896, Expediente 1.

Various. 1962. *Permisos de aprovechamiento forestal.* AGEO: Asuntos Agrarios, Serie V, Problemas Por Bosques, Legajo 897, Expediente 1.

Vasconcelos, José. 1929. Altiplanitis. *México Forestal Tomo* VII (3) (March): 41–43.

Vazquez León, Luis. 1986. *Antropologia politica de la comunidad indigena in Michoacán, coleccion cultural.* Morelia, Michoacán: Secretaría de Educacion en el Estado de Michoacán.

Velasco Pérez, Carlos. 1987. *Historia y leyenda de Atepec, Ixtlán, Oaxaca.* Oaxaca de Juárez, Mexico: Gobierno del Estado de Oaxaca, Dirección General de Educación, Cultura y Bienestar Social.

Vera, Manuel R. 1903. *La Dasonomía: Generalidades.* Mexico: Secretaría de Fomento.

Villela, F., and L. Samuel. 1997. De vientos, nubes, lluvias, arco iris: simbolización de los elementos naturales en el ritual agrícola de la Montaña de Guerrero (México). In *Antropología del clima en el mundo hispanoamericano*, tomo I. Ecuador: Ediciones Abya-Yala.

Wade, Peter. 1997. *Race and ethnicity in Latin America.* London: Pluto Press.

Wakild, Emily. 2007. Naturalizing modernity: Urban parks, public gardens, and drainage projects in Porfirian Mexico City. *Mexican Studies/Estudios Mexicanos* 23 (1):101–123.

Weiner, Tim. 2002. 16 arrested in killings of 26 over land disputes in Mexico. *New York Times* (June 3, 2002):3.

Whitmore, Thomas M., and B. L. Turner, II. 2001. *Cultivated landscapes of middle America on the eve of the conquest.* New York: Oxford University Press.

Williams, Raymond. 1983. Community. In *Keywords: A vocabulary of culture and society.* London: Oxford University Press.

Wolf, Eric. 1955. Types of Latin American peasantry: A preliminary discussion. *American Anthropologist* 57:452–471.

Wolf, Eric. 1957. Closed corporate peasant communities in Mesoamerica and Central Java. *Southwestern Journal of Anthropology* 13 (1):1–18.

Wolf, Eric. 1986. The vicissitudes of the closed corporate community. *American Ethnologist* 13 (2):325–329.

World Bank. 1997. *Staff apraisal report: Mexico community forestry project.* Washington, DC: World Bank Sector Leadership Group and Mexico Department, Latin America and the Caribbean Regional Office.

Worster, Donald. 1979. *Dust bowl: The southern plains in the 1930s.* New York: Oxford University Press.

Wynne, Brian. 2005. Reflexing complexity: Post-genomic knowledge and reductionist returns in public science. *Theory, Culture & Society* 22 (5):67–94.

Young, C. M. 1975. The social setting of migration: Factors affecting migration from a Sierra Zapotec Village in Oaxaca, Mexico. London: Ph.D., Anthropology, London University.

Young, C. M. 1982. The creation of a relative surplus population: A case study from Mexico. In *Women, development, and the sexual division of labor in rural societies*, ed. L. Beneria. New York: Praeger.

Zavattini, Cesare. 2005. *Dal soggetto alla sceneggiatura: Come si scrive un Capolavoro: Umberto D, belle storie.* Parma: Monte Universitá di Parma.

Zerecero, Anastasio. 1902. *Benito Juárez: Exposiciones (como se gobierna).* Mexico City: F. Vasquez.

Zorrilla Sangerman, Francisco. 1973. *Relacionado a un incendio forestal en Cerro Pelon, Ixtlán, Oaxaca.* Asuntos Agrarios: AGEO.

Index

Italicized page references indicate Spanish language, indigenous languages, or Latin names or titles (except for proper names).

Agrarian reform, 37, 97, 183. *See also* Ministries of Agriculture and Agrarian Reform; Reforms
Agricultural, 9, 16, 33, 36, 44, 46–47, 59
burning, 12, 13, 41, 49, 57
communities, 51 (*see also* Communities)
crops, 44, 67
development, 45
land, 36–37, 44–45 (*see also* Land)
practices, 16
Agriculture, 33, 85
burning, 105–107, 114–115, 117–119 (*see also* Agropastoral)
conceptual and institutional separation from forestry and forests, 43–46, 144–145
ecological zones in Sierra Juárez, 67–69
expansion and abandonment in the Sierra Juárez, 69–70, 80–81, 89-90, 121, 129, 135–144
fires, 105–107 (*see also* Agropastoral)
knowledge as a critique of modern forestry, 142–144
milpa, 118, 137–139, 141–142 (*see also Milpa)*
Ministry of, 44, 47, 109, 125, 171
subsistence, 90
swidden, 44, 106, 118, 120, 123, 137, 145, 210 (*see also* Swidden)
zapotec understandings of agriculture and forests, 98–99
Agropastoral
burning, 43, 46, 56, 58, 117–119, 153, 154–155
fires, 16, 154, 174, 195, 238
Alemán, Miguel, 113, 122
Anthropogenic forest, 89
Anthropologists, 7, 13, 24, 99, 162, 211, 239, 241
environmental, 4, 89
Anthropology, 11, 13, 23, 24, 179, 240
environmental, 26
Anti-concession movement, 132
Appropriation, 101, 140
Asamblea Comunal (Community Assembly), 111
Atepec, 106, 127–128, 131, 218
Audiences for performances of bureaucratic authority, 2–6, 11, 13–15, 17, 20–25, 101, 107, 109, 147–149, 151, 153, 155–157, 166, 169, 172, 174, 176, 200–202, 213, 221, 236–238, 240
rural, 107
urban, 41, 45, 57, 95, 106, 167–168, 173, 200, 212, 236 (*see also* Urban)

Authority
 community, 214, 220–221
 political and bureaucratic, 2, 4, 19–21, 96–98, 149–153, 167–169, 194–195, 199–200
 technical, 43, 57, 58–60, 110–111
 uncertain, 1, 4
Autonomy, 9–12, 27, 37, 58, 96, 142, 212, 214, 231
Azuela, Antonio, 160, 165, 169–170, 173, 176–177

Barad, Karen, 20
Barragan, Luis, 194
Biodiversity, 2, 7, 19, 21, 33, 69, 156, 171, 183, 185
 protection, 32, 34, 152, 163, 180, 227, 231
 protection institutions, 227, 231
Biologists (*biólogos*), 34, 54–55, 150, 162, 172, 186, 190
Bosque mesofilo, 66
Bracero migrant worker program, 80
Bribes (*mordidas*), 50, 104, 111, 183, 188, 225
Bureaucracy, 3, 7–9, 39, 50, 163–164, 167, 169, 186, 201, 206, 220
 authority, 21, 149
 careers, 158, 167
 control, 32, 95
 federal, 59, 203, 213
 indigenous forestry, 3, 220–223, 235–236
 Mexican forestry, 3
 as performance, 3 (*see also* Performances)
 transparency, 56 (*see also* Transparency)
Burning
 agricultural or pastoral, 12–13, 43–44, 75–78, 89–90, 105–107, 118–119, 210
 agricultural or pastoral, state regulation of, 46, 49, 56–57, 153–155, 157, 173–174, 192–195
 agricultural or pastoral burning, community regulation of, 139–140
 charcoal, 102–103
 documents, 170
 forests, 41, 83–89
 history of, 82
 use in war, 78–81

Campeche, 36
Campesinos, 52, 57
Capitalism, 8, 24
Carabias, Julia, 55, 162, 164–165, 169, 171–172, 174
Carbon, 1, 7, 19, 21, 32, 180, 236
Cárdenas, Alberto, 174
Cárdenas, Lázaro, 37–38, 41, 43–44, 50–51, 95–98
Cargos, 209, 213–219
 hierarchies, 213–214
 holders, 218–219
 offices, 214
 religious, 215
 system, 96, 213–214, 216–217, 230
Caudillos, 70, 73, 77, 91
 leaders, 91
CEMASREN, 59, 183
Change
 climate, 1–2, 19, 32, 39, 85, 95, 100–101, 115, 210, 231, 239
 environmental, 25–26, 61, 120, 129, 140, 231
Chapingo, 28, 33–34, 48, 54, 162–163, 186
Charcoal, 52, 74, 102, 104, 110–111, 114, 155, 194–196
 burners, 49, 105
 burning, 52, 99, 102–103, 130, 154
Chávez Morado, Jose, 17–18, 151. *See also Nube de mentiras*
Chiapas, 36, 205
Chihuahua, 36
Chinantec, 61, 94
 indigenous people, 122
Ciudadanos (citizens), 1, 31, 40, 98, 107–108, 202, 212–213, 216, 230, 237, 241
Civic epistemologies, 24, 202, 232, 237

Climate, 27, 31, 34, 40–41, 65–67, 88, 94, 99–101, 208
change, 1–2, 19, 32, 39, 85, 95, 100–101, 114–115, 210, 239
control, 105
Cochineal, 5
boom, 68, 70
cactus, 67–68
dye, 70, 90
dye production, 68
farmers, 70
fields, 90
insect, 63, 68
trade networks, 68–69
Coffee, 66, 69, 72, 123, 135, 187
cultivation, 69, 135
prices, 69
production, 69, 80, 186
Collusion, 3–4, 7, 12, 15, 21, 24, 155, 190, 197–199, 226, 236–238
Colonization, 90, 129
Comisariado de Bienes Comunales (Community Property Commission), 96, 139, 213, 216
Comisariado system, 97
Comisión del Papaloapan (Papaloapan Commission), 121–122, 140–141, 145
Commodity production, 68
Communities
authority, 53, 102, 214, 232
autonomy, 37, 21
elders, 9, 29
forestry businesses, 53–54, 110–111, 139, 142, 144, 206, 209, 216
identity, 11, 121
of Ixtlán, 5–9, 11, 29, 53, 85–86, 97, 100, 106, 130, 155, 203, 211, 219
political power, 11, 120, 144
politics, 95, 209–210, 212, 218–219, 230, 232
representatives, 102, 126, 147–148, 155, 157
Comuneros (commoners), 78, 97, 106, 121, 127–129, 138–139, 141–142, 144–145, 149, 187, 208–209, 212–213, 216–222, 225, 228–230, 232
Comunidades (communities), 213, 216
Agrarias, 213, 216 (*see also* Ixtlán)
CONAFOR (National Forestry Commission), 47, 158–159, 168, 174
Confianza (confidence) employees, 164, 181–182
Conservation, 2, 4, 19, 21, 28, 33–34, 38, 46, 145, 163, 185–187, 190, 199, 236, 240
officials, 4, 240
Conservationists, 7, 156, 171, 187–188, 240
Conzatti, Cassiano, 76–77, 89
Cooperatives, 38, 43, 50–51, 103, 111
Corporaciones de Defensa contra Incendios (Community Fire Defense Corporations), 105
Corruption, 2, 7, 14, 55–56, 60, 111, 158, 165–166, 170, 187–189, 198–200, 236
accusations of, 156, 163, 165–166, 185, 187–191, 195, 198, 200–201
Crops, 39, 44, 62, 66–67, 99, 122–123

Dams, 4, 13, 122–123
De base (base) employees (*empleados de base*), 164, 182
Deforestation, 31, 35, 40, 75–76, 89, 95, 100–101, 106, 114, 122–123, 129, 141, 174
de la Fuente, Julio, 99–101
Demographic collapse, 68
Dendrochronology, 85
Departamento Forestal y de Caza y Pesca (Department of Forest, Hunting, and Fishing), 39, 41
de Quevedo, Miguel Angel, 40, 42, 48, 76, 89, 103, 109, 159, 196

De Sica, Vittorio, 3
Desiccation theory, 34, 40, 45–46, 76–77, 89, 101, 103, 223
Development as state project
 and depoliticization, 13, 22–24, 148, 186, 214, 241
 development NGOs in Oaxaca, 186–187
 and industrial forestry, 125–126
 of natural resources, 41–47, 121–124
Día del árbol (Tree Day), 101–102, 107–109
Díaz, Porfirio, 36, 38, 70, 72–73, 75, 77
Discourses, 14, 40, 46, 120, 141, 232
 of antipolitical development, 23–24
 of desiccation, 39–42
 of environmental degradation, 12, 16, 31, 39–44, 58–60, 115, 144–145, 201–202
 of indigenous isolation, 119
 official, 12, 14, 41, 60, 118, 140, 144, 157, 173–176, 201, 240
 stable, 59
 as storytelling about corruption, 187
 of swidden agriculture, 41–44
Dissent, 152, 156, 166, 172, 198–199, 212
Documentadores (documenters), 225–226
Domínguez, Aldo, 149–150, 152–155, 162–163, 194, 198–199, 201
Drugs, 152
 traffickers, 150
Durango, 36

Echeverría presidency, 38, 52
Ecologies, 6, 11, 27, 34, 45, 55, 62, 64–65, 88, 91, 143, 179, 185, 187, 211, 222
 ecological zones, 65, 69, 220
Economies, 38–39, 74–75, 78, 80–81, 91, 119, 131, 135, 149, 182
 economic crisis, 33–34, 38, 47, 53, 55, 163, 170, 183, 185
 Mexican, 33, 80
 political, 61, 63–65, 86, 91, 119, 134
 projects, 7
Ejidos, 37, 53
 forestry businesses, 52
 unions of, 52
Elites
 regional political, 95, 112
 rural political, 96
 state political, 113, 125
 urban, 34
Empleados de confianza (confidence employees), 164, 181–182
Environment
 activism, 12, 190, 198, 212
 advocacy, 39
 changes, 25–26, 61, 120, 129, 140, 231
 crisis, 32
 degradation, 25, 31, 36, 40, 77, 118–120, 123, 126, 129, 133, 141, 148, 154–155, 199, 201, 210
 discourses, 12, 16, 31, 58, 201–202
 histories, 26, 32, 37, 39, 61, 64–65, 91, 98, 140
 NGOs, 53, 168, 186
 order, 37, 115, 118
 policies, 31, 39, 167, 172
 politics, 33, 92, 95, 120, 132
 problems, 32 (*see also* Problems)
 restoration, 41, 94–98, 115, 119, 168
 science, 31, 99, 114
 of the Sierra Juárez, 65, 90
 theories, 16, 39, 45, 93, 95, 99, 101–102, 104, 114, 121, 140
Environmentalists, 54, 133–134, 141, 150, 157, 175, 179, 229, 231
 Mexican, 34
Epistemic community, 12
Ethnography, 11, 27–28, 99–100
Expertise, 13, 23–24, 153, 171, 175, 202, 232
 scientific and technical, 11, 13, 22

FAO (Food and Agriculture Organization of the United Nations), 11, 37, 40, 196–197
FAPATUX (*Fábricas Papeleras de Tuxtepec*/ Tuxtepec Paper Factories), 8–9, 113, 115, 117–134, 136–137, 140–141, 143, 208, 213–214, 226
Farmers, 3, 21, 25, 27, 40, 42, 49, 56–57, 60, 65, 67–68, 70, 91, 105, 107, 122–123, 135, 143–144, 160, 174, 192, 238
  migrant swidden, 118, 140 (*see also* Swidden)
Ferguson, James, 13, 22–23, 175, 201, 241
Fertilizer, 44, 84, 155
Fires, 27, 46, 57–58, 64, 67, 75–78, 80, 82–86, 88–89, 91, 99, 103, 105–107, 129, 143, 157, 195, 216
  agropastoral, 16, 44, 64, 74, 77, 89, 92, 138, 154, 174, 195, 238
  ecology study, 27
  fighting, 101, 105, 136, 138–139, 144, 171
  forest, 9–10, 16, 31, 39, 41, 43–44, 46, 82, 89, 94–95, 102, 105–106, 117, 129, 137, 140–141, 143, 145, 149, 153–155, 157, 174
  history studies, 64, 85
  of wars, 61, 77, 80 (*see also* Wars)
Firewood, 36, 49, 52, 74, 84, 102, 104, 110–112, 114–115, 130, 153–156, 172, 181, 191–192, 194–197, 201–202, 216
  cutters, 103–104, 112–114, 192
  cutting, 52, 103–104, 107, 153–154, 157, 171, 194–195, 200–201
  regulations, 154, 156, 171, 173, 175, 192, 194–195, 200, 202
  sellers, 104, 107, 114
Fir species, 67. *See also* Trees
Floods, 34, 39–40, 45, 89, 99, 121–123, 145
  protection, 39
Fondo Nacional de Fomento Ejidal (National Ejidal Development Fund, FONAFE), 125, 130
Forestry
  bureaucracies, 3, 5, 7–8, 11, 25, 61, 93, 177, 213, 236
  careers, 54
  community, 9, 38, 51–54, 64, 74, 110–111, 117, 119, 130, 133, 140, 147–149, 154, 206, 209–212, 220, 227–228
  convention, 148, 162, 174, 199, 228, 231
  cooperatives, 43, 50–51
  industrial, 6–11, 33–34, 52, 92, 117, 120, 124, 137–138, 142–144, 147, 156–157, 175, 201, 228, 232
  institutions, 2, 5–6, 11, 32, 34, 59–60, 145, 157–158, 173, 182, 203–204, 236
  Mexican, 3, 11–12, 32, 45, 51, 143, 164, 235 (*see also* Mexico)
  modern, 142–143
  officials, 10–11, 15–20, 49–52, 55–59, 101–106, 112–114, 148–177, 180–188, 191–202, 235–236, 240
  police, 110
  practices, 110, 113
  regulations, 11–12, 14, 16, 32, 34, 44, 50, 56, 93–94, 101–102, 104, 108, 147, 152, 160, 163, 179, 182
  school (*Escuela Nacional Forestal*), 48, 163
  science of, 2, 5–6, 9, 25, 31–34, 37, 93–94, 99, 115, 118, 124–125, 203, 221, 232, 235, 239–240 (*see also* Silviculture)
  technicians, 9–10, 86, 110, 124, 128, 134, 142, 185, 203–204, 222–223, 232
Forests
  administration, 39
  burning, 41, 44

Forests (cont.)
communities, 5–12, 28–29, 37, 51–54, 119, 127–134, 181, 185–187, 228–231
ecology, 11, 27, 45, 62, 64, 91, 211
fires, 9–10, 16, 31, 39, 41, 44, 46, 82, 89, 94, 105–111, 113–114, 129, 137, 140, 145, 152, 154, 157, 174, 181
industrialization, 34, 50, 126, 207, 217
laws, 38–39, 41, 46, 53, 55–56, 58, 94, 134, 172
management, 7–11, 36, 42, 46, 52, 55, 58–59, 73, 81, 84, 94–95, 119, 134, 142–143, 149–150, 176, 180, 182, 185, 190, 206–209, 219–220, 226, 231 (*see also* Management)
pillager (*rapamontes*), 111
pine and pine-oak, 2, 35–36, 38, 44, 51, 58, 61–62, 67, 69, 77, 80, 82, 84, 86, 89, 93, 118, 124, 129, 144, 150, 203–204
protection, 40–42, 73, 121, 176, 227
regulations, 39, 104
service, 5–6, 13, 16, 25, 28, 32, 34, 39, 43–44, 46–48, 50–51, 54–61, 86, 89, 92, 94, 99, 101–106, 111, 113, 117–118, 120, 126–127, 132, 134, 139, 141, 144, 147, 149, 153, 157, 163, 196, 202, 214, 221, 225, 227, 235–237 (*see also* Mexico)
subsidies, 53
workers, 6, 139, 204, 218
Foucault, Michel, 16, 20
Fox administration, 164–165
French Intervention, 70
Funcionario de carrera, 164

Gangsters, 93
García, Antonio, 219, 221
Garizurieta, César, 167
Gender roles, 217
Gomez, Jorge, 196–197
Governments
federal, 39, 94, 107, 113, 118, 123, 127, 131, 214, 217, 222, 229
foresters, 9–10, 60, 110, 126, 223
institutions, 6, 121, 164, 166, 173
local, 97
officials, 6–7, 11–12, 14, 16, 23, 53, 81, 104, 118, 132, 150, 162, 166, 169, 187–188, 221–222
state, 106, 119
subsidies, 150, 182
technicians, 31
Gremio, 54
Guelatao, 68, 70, 75–76, 96, 126, 132, 208
Gulf of Mexico, 32, 65
Gupta, Akhil, 143, 179

Haraway, Donna, 20, 24
Herzfeld, Michael, 15, 156, 160
Histories, 8, 20, 25, 28, 31–33, 36, 61–65, 72, 77–78, 81–82, 85, 88, 90, 92, 96, 98, 127, 134, 140–141, 143, 151, 158, 162, 208, 219, 235, 237, 241
ecological and political, 62
environmental, 26, 39, 61, 65, 91, 98, 140
of fires, 64, 67, 86, 88, 91, 143
of floods, 39
of forestry, 19, 31–32, 118, 145
of institutions, 32
of Mexican forests, 34, 164
political, 6, 90–91
Human, 16–17, 20, 24–27, 60, 68, 80, 82, 85, 89, 91, 99–100, 122, 166, 168, 180, 199, 211, 221, 232
agency, 25–27
history, 88
settlements, 63, 65, 67–68, 90
Hybridization, 140, 143

Identities, 9, 11, 15, 69, 75, 119, 121, 133–134, 139, 142, 144, 168, 212
Ideologies, 10, 100, 114, 118, 124, 140, 142, 168
official, 102, 106, 113, 121, 136, 138, 141, 201

Ignorance, 6–7, 19, 41–42, 60, 107, 138, 147, 154, 163, 166, 170, 176, 189, 195, 200, 203, 237–238
production of knowledge and, 11, 13, 201
official, 12–13, 17, 59, 176, 192, 199, 201–202
rural, 41, 60
Imagination, 32, 57, 60, 201, 235, 241
power of, 242
Immigrants, 204, 207–208
Immigration, 207
India, 31
Indigeneity, 41, 73, 133–134, 143, 211–213
Indigenous, 5–8, 33, 36, 41, 56, 61, 68, 72–73, 94, 107, 117, 119, 121, 133, 143, 162, 204, 211–213
agriculturalists and pastoralists, 119
bureaucracy, 3, 7
communities, 5–9, 25, 60–62, 68, 70, 72, 77, 81, 89, 93, 95, 113, 118–119, 121, 126, 147, 189, 202–204, 212, 235
community of Ixtlán de Juárez, 5, 7, 11
forest communities, 6, 29, 37, 60–61, 235, 239
forest dwellers, 32
identity, 75, 121, 133–134
industrial, 203
languages, 74
peasant agriculturalists, 34
people, 2, 5–6, 11, 13, 37, 41, 44, 52, 60–61, 64, 73, 75, 92, 95–96, 99, 117–118, 120, 122–123, 134, 143, 149, 157, 162, 203, 212, 235, 242
political institutions, 63
populations, 68 (*see also* Populations)
Indigenousness, 8, 121, 133–134, 147
Industrialization, 34, 38, 50, 52, 98, 126
Industrial logging units (*Unidades Industriales de Explotación Forestal* [UIEF]), 38, 51, 113, 125
Inequality, 212
Institute of Ecology (IEE), 34, 55
Institutional Revolutionary Party (*Partido Revolucionario Institucional* [PRI]), 36, 38, 96, 164–165
Instituto de Reforma Agraria (Agrarian Reform Institute [IRA]), 97
Instituto Nacional Indigenista (National Indigenous Institute [INI]), 121
International trade, 68, 70
Ixtepeji, 72, 78, 131
Ixtlán (de Juárez)
community of, 5–9, 11, 29, 53, 85–86, 97, 100, 106, 126, 130, 155, 203, 211, 219
comunidad agraria of, 216
Ixtlecos, 96, 207–208

Jalisco, 52, 127
Jasanoff, Sheila, 20, 23–24, 145, 148, 202, 227, 232, 237, 239
Juárez, Benito, 70, 75, 96, 126

Knowledge
agroecological, 142
authoritative, 59, 237
bureaucratic, 4, 201, 220, 225
ecological, 54, 60, 78, 92, 209
forms of, 2, 4–5, 8–9, 12, 25–26, 148, 173, 175–176, 204, 227, 232, 235–237, 240–241
ignorance and, 11, 19 (*see also* Ignorance)
local, 4, 9, 13, 56, 145, 221–222, 238
local ecological, 56
making/making of, 4, 7, 13, 15, 19, 21, 23–26, 92, 140, 149, 158, 175–176, 199–203, 208, 221, 227, 231–232, 237–238, 240–241
modern economies of, 61
official, 6–16, 19–20, 22–25, 50, 58–61, 101–103, 148–153, 155–158, 170–176, 180–185, 192–196, 199–202, 227–231

Knowledge (cont.)
official environmental, 43, 95
as performance and practice, 26
popular, 12, 145
production of, 11, 200
projects of, 16, 64, 120, 236, 239
public, 4, 17, 19, 24, 110, 141, 148, 151–153, 157, 175–176, 236–237
scientific, 22–23, 32, 41, 43, 60, 81, 175–176, 180, 202, 223, 232–233, 236
social science of, 240
technical, 11, 32, 46, 60, 148, 152–153, 200, 209, 211, 219–222, 230–232
transparent, 2, 15, 19–21, 152, 229, 236

Labor, 52, 68, 71, 77–78, 80, 91, 97, 115, 126, 128, 130, 134, 217
relations, 118
unions, 17
Laborers, 135
migrant, 135
Land
communal, 37, 72–74, 96–97, 105–106, 111, 183, 208, 230
degraded, 42, 141–142
ownership, 63, 70, 72, 77, 95–97, 114, 122, 127, 139, 192
reform, 32, 36–37, 45, 50, 52
use change, 90–92, 135–140, 181–182
Landowners, 51, 54–55, 72, 77, 199
Landscapes, 5–6, 14, 20, 25, 35, 39–40, 55, 59–60, 61, 63–65, 74–78, 81, 86, 89–90, 92, 98, 105, 109, 112, 124, 130, 140, 145, 162, 179, 195, 225
change, 63
changing, 91, 118
forest/forested, 60, 89–90, 92
military, 70
Latour, Bruno, 23, 26, 157, 172, 199–200, 241
LEGEEPA (General Law of Ecological Equilibrium and Protection of the Environment), 54, 172
Legibility, 6, 15, 19, 61, 73, 94, 192, 200–201
Liguori, Francisco, 167
Local
contexts, 4, 6, 10, 13, 229, 238–239
details, 6, 13, 31, 241
Loggers, 6–7, 51, 57, 93–94, 111–113, 118–120, 127–128, 203–204, 221–226
community, 9, 16, 111, 128, 221–226, 227, 229
wildcat, 93, 107, 117
Logging
clear-cut, 45, 126–127
companies, 6, 8–9, 36, 38, 50–55, 80, 94–95, 110–115, 117–120, 123–125, 129–141, 181, 190, 213–214
concessions, 6, 38, 51–53, 55, 117, 124–125, 128–129, 131, 133, 187
contractors, 54
contracts, 111
forest, 2, 220, 232
illegal, 7, 12, 95, 111–112, 151, 153–155, 231
industrial, 37, 45, 69, 98–100, 112–113, 120, 126–127, 130
permits, 49, 110, 113, 125, 152, 155, 175
practices, 11, 16, 45, 57, 103, 120, 128–129, 210–211
regulations, 49, 56, 59, 94, 101, 107, 117
selective, 45, 58–59
technologies, 118, 127–128
López, Araceli, 190
López Portillo, Jose, 122, 132
Luna, Jaime, 132–133

Management
plans, 8, 42, 49–55, 58–59, 73, 84, 124, 134, 144, 149–150, 176, 179–183, 185–186, 188, 191, 195, 198, 206, 219–222, 226–227, 229, 231–232
practices, 45, 50, 142
Mapping, 2, 19, 33, 72–73, 119
Mares, Jose, 165–168, 201

Markets, 8, 21, 52, 62, 64–65, 70, 102, 104, 110, 114, 134–135, 155, 187, 194, 223
carbon, 7, 19, 21, 180, 236
environmental, 240
financial, 1, 19, 21
world, 69
Memory, 36, 86, 99–100, 124, 144–145
Mestizo military/political leaders, 70
Mestizo political entrepreneurs, 70
Método de Desarrollo Silvícola (MDS), 45
Mexico
anthropologists, 99, 162
environmentalism, 39
forestry, 2–4, 11–12, 14, 28, 32–33, 45, 51, 60, 143, 157–158, 160, 164, 177, 200–201, 235–237, 240
forests, 15, 17, 19, 21, 28, 34, 36, 46, 112, 202, 235
forest service, 12, 28, 32, 59, 64, 92, 94, 143, 147, 235–236
Mexico City, 28–29, 33–35, 39–40, 50–51, 53, 56–57, 110, 159–164, 170–174
oil industry, 96
political culture, 10, 202
Revolution, 6, 36–38, 69
society, 121, 148, 162
state, 32–34, 36–37, 44–50, 60–61, 94–95, 117–118, 150–153, 200–202, 214
valley, 39–40, 42, 59–60, 75
Michoacán, 36, 52, 110, 127
Community of San Juan Nuevo Parangaricutiro in, 53
Militias, 70, 90
Followers, 91
Zapotec, 73, 91
Milpa, 44, 138–139, 142, 144, 155
agriculture, 118, 137–139, 141–142
cultivation, 121, 142–143
fields, 135
Mineral, 45, 57, 63, 98, 112, 152
exploitation, 119
Mines, 1, 73–74, 81
La Natividad, 74, 80, 135
silver, 90
Mining, 5, 51, 62, 69–70, 72–73, 75, 90–91, 94. *See also* Oaxaca
boom, 73–74, 90, 98
centers, 70, 74, 76, 90, 121
economy, 75, 78
silver and gold, 69–70
towns, 80
Ministries of Agriculture and Agrarian Reform, 44, 171
Modernity, 8, 14, 121, 143, 158, 207, 235
Monreal Rangel, Saul, 163–164
Mosse, David, 158, 237
Movements, 23, 96–97, 119, 132–133, 162, 212
popular, 6, 8
Municipalities (*Municipios*), 203, 205, 213, 215, 217
officials, 213
rural, 77

National
events, 77, 119
forestry laws and policies, 5
forestry statistics, 158, 170, 203–204, 225
parks, 51, 240
politics, 69–70, 90, 134, 214
regimes, 77
rituals, 101, 109
statistics, 3, 31, 50, 183, 197, 203, 232, 238
National Autonomous University of Mexico City (UNAM), 34, 54
National Institute of Ecology, 34
Nationhood, 75
Natural resources, 31, 47, 53, 92, 123, 182
Nature
constructions of, 25
and culture, 43–46
culture, and human agency, 25–28, 144
protection, 31, 33, 179

Nature (cont.)
resisting human intentions, 16, 25–28, 64–65, 88–91, 120, 221–222
rituals, 98–100, 107
and society, 8, 13–14, 171–173
spirits, 100
Neoliberal, 4, 239–240
economics, 19
reform, 7, 53
Nongovernmental organizations (NGOs), 53–54, 132–134, 147, 168, 185–187, 190–191
comunalidad, 132
movement, 132
workers, 186
Nube de mentiras (The Cloud of Lies), 18, 151. *See also* Chávez Morado, Jose
Nuijten, Monique, 151, 212
Numbers, rhetoric of, 149–151

Oak, 22, 52, 63, 65, 67, 69, 74, 81–85, 89, 91, 103, 112, 115, 128–130, 139, 204, 223. *See also* Trees
Oaxaca
city of, 5–6, 66, 76, 89, 93, 102–104, 121, 147, 179, 181–182, 187, 204, 228
Compañía Forestal de, 94, 113, 117, 124, 132
Sierra Juárez, 5, 7, 14, 25, 53, 61 (*see also* Sierra Juárez, environment of)
Sierra Sur of, 93
state of, 3, 5, 11, 66, 102, 130, 147, 149, 153, 163, 181–182, 191, 235
valley, 62–63, 65, 68–69, 72, 104, 114, 150
ODRENASIJ, 132–134
Officials. *See* Forestry, officials

PAN (National Action Party [*Partido de Acción Nacional*]), 38, 164–165
Parastatal, 6, 9
control, 58
timber corporations, 52
Partido Revolucionario Institucional (PRI), 36, 38, 96, 164–165
Pastoral, 13, 74–76, 89, 105, 118, 192
Pastoralism, 49, 81, 85, 117–118, 121, 137
Pastoralists, 49, 70, 77, 91, 105, 118–119, 144, 174
Peasants, 3, 34, 37, 41, 44, 57, 81, 84, 90–91, 106, 133–134, 163, 236
Pérez, Zenaido, 138–140, 155
Pérez García, Rosendo, 72–73, 77, 80, 97, 122
Performances
bureaucratic, 112, 115, 148
official, 10–11, 24, 149, 157, 175, 200, 218
of public reason (and rural ignorance), 20, 41
Political
action, 6, 60
affiliations, 91
autonomy, 11, 212
centralization, 109
concerns, 91
conflicts, 128, 212
crisis, 34
culture, 2, 10, 24, 32, 59–60, 200, 202
and epistemic work, 21, 176, 188–189
events, 38, 92, 119, 151
institutions, 63, 81, 211, 214
life, 59, 165
logics, 91
metaphors, 20
theory, 19–20
Politicians, 8, 55, 57, 75, 90, 95, 125, 167–168, 172, 174, 189, 200, 228, 236–237
Politics, 12–13, 23–24, 69–70, 90–92, 119–120, 175–176, 209–210, 212–214, 218–219, 227–228, 230–232
Populations, 36–37, 44, 68–69, 80, 90, 117, 135, 222
human, 68, 85
indigenous, 68

Porfiriato, 70, 98
Power
 institutional, 5, 21, 157, 175, 188
 and knowledge, 14, 16, 151, 210
 material, 39, 153, 166
 of publics, 2, 7, 21
Pozo, Jose, 137, 139, 142, 209–211, 227, 231
Practices. *See also* Burning; Logging
 daily, 12–14, 16, 166, 195
 of paperwork, 49–52
 state, 10, 195
 theories of knowledge as performance and, 26
PRD (Party of the Democratic Revolution [*Partido Revolucionario Democrático*]), 164, 172
Presidente de bienes comunales (President of community property), 205, 216, 219
Private control, 58
Privatization, 72
Problems, 25, 105, 113, 165, 185, 217, 226
 technical and environmental, 32
PROCEDE, 229–230
PROCYMAF project, 147, 154, 182
PROFEPA (environmental prosecutor's office), 155–156, 162, 173, 182, 192, 194–195

Ramírez García, Luis, 78
Ranz, Miguel and Jose, 93, 111
Reforms, 28, 32, 37–38, 44–47, 50, 52–53, 56, 68, 70, 97, 166, 172, 183, 239
 wars of, 70
Regulations
 forestry, 11–12, 16, 32, 34, 39, 44, 50, 56, 93–94, 101–102, 104, 108, 152, 160, 163, 179, 182
 individual, 59
Representations, 25–26, 40–41, 120, 148–150, 156–159, 165–166, 171–175, 227–228
República de indios, 68
Resistance, 14–16, 39, 61, 64–65, 81, 86, 88, 90–92, 120, 126, 132, 140, 175, 191, 221–222, 232
Restitución (restitution), 36
Revolution, 6, 32, 36–38, 48, 69, 73, 90, 98, 112. *See also* Mexico, Revolution
 in the Sierra, 77
Rhetoric, 13, 19, 101, 107, 123, 149–150, 152–153, 157, 166–167
 of fire, 153
 official, 13, 167
 of transparency, 192 (*see also* Transparency)
Rio Grande, 67–68
Rivera, Diego, 33
Roads, building of, 4, 11, 40, 94, 119, 122–124, 126
Romero, Francisco, 163, 170–171, 201
Rosas, Juan, 190, 198
Rozas, 107, 154. *See also* Swidden
Ruiz, Gustavo, 187, 191

San Felipe del Agua, 102, 104
San Luis Potosí, 74
San Pablo Etla, 102–103
SARH (*Secretaría de Agricultura y Recursos Hidraúlicos* [Ministry of Agriculture and Hydraulic Resources]), 47, 59, 171, 183, 197
Sawmills, 53, 187
Scott, James, 12–15, 19, 122, 200–201, 229, 241
Secularization, 97, 107
SEDUE (Ministry of Environment), 171
"Seeing Like a State," 15, 19, 122
SEMARNAP (*Secretaría de Medio Ambiente y Recursos Naturales y Pescas* [Ministry of Environment, Natural Resources and Fisheries]), 29, 47, 53, 56, 147, 154, 156, 158–165, 167, 171–174, 177, 179–183, 185–188, 190–195, 197–199, 219–220, 227, 230–231

SEMARNAT (*Secretaría de Medio Ambiente y Recursos Naturales* [Secretariat of Environment and Natural Resources]), 47, 167, 185, 196, 199, 214
Serranos, 69–70, 77–78, 90–91, 99, 101, 105, 112, 119–120, 133, 228
Servicios técnicos, 179, 183, 185–187, 190, 195, 198–199, 209–210
 prestadores de, 186
Sierra Juárez, environment of, 65–68
Sierra Madre Oriental, 37, 65
Sierra Sur, 93, 124, 132, 181
Silencings, 4, 13, 153, 157–158, 175–177, 179, 194, 199, 236
 practices of, 180
Silvicultural, 117
 science, 43, 46, 60
 systems, 45, 57–58
 treatments, 57
Silviculture, 31, 45, 48, 57, 59–60, 118
 scientific, 34, 39
Soberanista leaders, 96, 98
Soriano, Juan, 191–192
Spanish, 68, 93, 121, 147
 conquest, 36, 68
 crown, 36
Stand dynamics, 64
 theories of, 81
State
 agencies, 101–102, 181
 anthropology of, 179
 authoritarian, 2, 14, 241
 authority of the, 6, 32, 152, 175, 188, 221, 237–238
 bureaucracies, 7–8, 9, 61, 63, 124
 and communities, 9, 132, 141, 214, 221, 227, 229
 criticism of, 171
 elites, 112–113, 125
 forestry, 6–8, 81, 92, 102, 145, 204, 232
 formation, 15
 ideologies, 10, 15, 114, 138, 140
 institutions, 8–9, 14–16, 31, 34, 44, 60, 95, 102, 114, 143, 145, 202–204, 213–214, 229, 236, 239, 240
 interventions, 10, 37, 52, 63, 145, 181
 knowledge, 9–10, 41, 64, 120, 149, 152, 174, 192, 202–203, 225
 legibility, 73, 94, 198, 200, 232
 legitimacy, 34, 172, 176, 203, 237–238
 Mexican (*see* Mexico, state)
 modern, 3, 13–15, 19, 41, 89, 126
 performing the/performances of, 15, 17, 19, 24–25, 42–43, 50, 108, 115, 148, 153, 188, 200, 235
 popular understandings of the, 150
 power, 5–6, 10–14, 39, 61, 151–152, 156–157, 171, 176, 179, 189, 220
 projects, 25, 34, 39, 41, 72, 92, 94, 96–97, 99, 115, 123, 232, 236
 rituals, 95, 107–108
 science, 5, 31
 "Seeing Like a State," 15, 19, 122
State-making, 5, 10, 32, 60, 148, 199, 200, 237
Statistical, 64, 86, 88, 150, 152, 153–154,156, 194, 197
 knowledge, 149, 153
 national reports, 31
Statistics, 3, 150–151, 154, 195, 201, 238
 national forestry, 153, 158, 170, 197, 203–204, 225
 official, 151, 171, 229, 241
 timber production, 50, 183, 232
Swidden, 44, 57, 80, 95, 106, 117–118, 120, 122, 137, 140, 143, 145, 154, 174, 210

Tamayo, Adulfo C., 112
Tamayo, Jorge L., 122, 124, 131
Taxes, 2, 3, 94, 113, 183, 228
 evasion, 93–94, 113, 117
 import, 53
Taxation, 20
Technology, 7, 74, 91, 111
 agroecological, 44
 logging, 127–128
 studies, 20, 22–23, 157, 201, 231

Texcoco, 33, 48, 221
Timber, 2, 10, 36, 49, 53, 67, 74, 84, 129, 143–144, 150, 165, 182–183, 187–190, 198, 208, 210, 220, 222, 225–228
 commercial, 37, 58
 companies, 51, 117, 123, 137
 concessions, 55, 113, 118, 126
 contracts, 93, 97
 corporations, 52
 cutting, 10, 37, 51, 110, 125, 130, 182, 220, 228–229
 extraction, 52, 115, 126–127, 144, 181
 industrial, 37, 130, 155–156
 pine, 37, 52, 110, 130, 137–138, 220
 poaching, 54
 prices, 131, 134
 processing, 51–52
 production, 31, 34, 37, 43, 69, 120, 131, 148–149, 153, 170, 176, 183, 222, 229, 232
 production statistics, 50
 smuggling, 112
 transport, 59, 128, 221
 transport documents, 49–50, 55, 111, 119, 188, 229
 transport regulations, 41
 tropical hardwood, 37
 volumes of, 49, 51–52, 110, 112, 125, 128, 141–142, 183, 185, 229
Tlacuache, 167
Transparency, 1–2, 17, 19, 21–22, 177, 179, 191–192, 233. *See also* Knowledge, transparent
 bureaucratic, 56
 documentary, 59
 official, 56–57, 158, 225
Transparency International, 2
Trees
 Abies hickelii, 67
 Abies oaxacana, 67
 ancient, 33, 129
 apricot, 209
 burning, 64, 83, 85, 143
 counting, 185, 203–204, 229
 cover, 57
 cutters, 206
 cutting, 8–9, 12, 34, 45, 57, 74, 94, 98, 103, 110, 113, 128, 139–140, 173, 201, 203, 220, 222, 225–226, 229
 Tree Day (*Día del Árbol*), 101–102, 107–109
 diseased, 225, 227, 232
 evergreen broadleaf, 66
 fir, 61–62
 fruit, 97–98, 107, 109, 123, 204
 jacaranda, 179
 line, 65
 liveliness, 86, 90–91
 living, 115, 120
 marking/marked, 115, 204, 220, 225–226
 materiality, 64
 measure, 86, 203
 nurseries, 107, 159, 182
 oak, 58, 65, 74, 81–85, 103, 115
 P. ayacahuite, 67
 P. douglasiana, 67
 P. michoacana, 67
 P. oaxacana, 67
 P. pseudostrobus, 67
 peach, 209
 pine, 25, 27, 58, 64–65, 67, 81–86, 88, 90, 92, 115, 129, 139, 140, 183, 198, 220
 Pinus patula (*pino rojo*), 67
 Pinus rudis, 67
 Pinus teocote, 67
 planters, 160
 planting, 8, 95, 98, 107, 109–110, 142, 168, 171, 173, 209, 210–211
 protect, 96, 107
 removing, 188
 resistance, 81, 83, 92
 rings, 86
 samples, 33, 85
 scarred, 84
 seedlings, 59, 75, 81, 168, 210
 shade, 109
 species, 185, 222, 225
 standing, 45, 58, 81, 103, 114

Trees (cont.)
timber, 67
treatment, 220
unmarked, 8
unruliness, 91, 129, 140, 222
younger, 82, 114–115, 117, 127, 140
Tsing, Anna, 200, 227, 239
Tuxtepec, 117, 124, 126, 131, 135
Valley of, 122, 124

UCFAS (*Unidad Comunitaria Forestal Agrícola y de Servicios* [Community Forestry, Agricultura and Services Unit]), 206, 217
UIEF, 38, 51, 125
Umberto D., 1
Uncertain authority, 1, 4
United States, 31, 33, 35, 69, 80, 135–136, 207–208, 215
University of Chapingo, 28, 33–34, 48, 54, 162–163, 186
Urban, 41, 102, 104, 115, 136, 147
audiences, 41, 45, 57, 95, 106, 167–168, 173, 200, 212, 236
centers, 102, 123
conceptions, 102
dress, 208
elites, 34
landscape, 162

Valle Nacional, 62, 66, 123–124
Velasco, Jose María, 75–76
Veracruz, 121, 135

Wars, 73, 78, 80, 92, 150
civil, 38, 69, 90
fires of, 61, 77, 80
French Intervention, 70
of independence, 68, 70
of the nineteenth century, 73
of Reform, 70
warfare, 5, 39, 62, 64, 68, 80–81, 90–91, 94, 119
World War II, 51, 80
Watershed, 39, 41–42, 51, 57, 121, 130, 227
control, 117
Papaloapan, 121–123
protection projects, 39, 59, 122, 227
Williams, Raymond, 211
World Bank, 147–148, 153, 186, 211, 227
forestry projects, 147, 182, 231
funding, 227

Yale School of Forestry, 28, 163

Zacatecas, 74
Zapatista rebellion, 134, 205, 212
Zapotec, 61, 72, 94, 98–99, 208, 212, 258
indigenous community of Ixtlán de Juárez, 5, 11
indigenous militias, 73, 91

Printed in the United States
by Baker & Taylor Publisher Services